UNIVERSITY OF STRATHCLYDE
30125 00733287 6

D1380762

PROCESS MANAGEMENT

Process Management

Why Project Management Fails in Complex Decision Making Processes

by

Hans de Bruijn
Delft University of Technology, The Netherlands

Ernst ten Heuvelhof
Delft University of Technology, and Erasmus University, Rotterdam, The Netherlands

and

Roel in 't Veld
Netherlands School of Government, Utrecht and University of Amsterdam, The Netherlands

KLUWER ACADEMIC PUBLISHERS
BOSTON / DORDRECHT / LONDON

A C.I.P. Catalogue record for this book is available from the Library of Congress.

ISBN 1-4020-2680-3

Published by Kluwer Academic Publishers,
P.O. Box 17, 3300 AA Dordrecht, The Netherlands.

Sold and distributed in North, Central and South America
by Kluwer Academic Publishers,
101 Philip Drive, Norwell, MA 02061, U.S.A.

In all other countries, sold and distributed
by Kluwer Academic Publishers,
P.O. Box 322, 3300 AH Dordrecht, The Netherlands.

Printed on acid-free paper

Printed in the Netherlands.

PREFACE

Managers, of public organizations as well as of large companies, increasingly face a difficult dilemma. One the one hand, much is expected of the decision making by governments and companies. The public want governments to provide safety, quality education, a proper infrastructure and accessible health care, for example. Company staff expect their managers to have a vision and ensure both innovation and continuity. On the other hand, possibilities for a government or corporate management to provide steering are limited. This is because organizations no longer lend themselves to hierarchical steering. Problems are so complex that the boards of many organizations lack the expertise to make the right decisions. This, then, is the dilemma: possibilities for steering are declining, while the demand for it is continuing undiminished.

In this book, we set out how to deal with this tension. How can politicians, CEOs or managers in public or private organizations provide steering, if they are unable to rely on hierarchy or on their expertise? The answer is: by designing and managing the process of decision making intelligently. The book presents a comprehensive description of this process design and process management.

Currently, attention for process design and process management is emerging everywhere, in Europe as well as in the United States, both in practice and in decision-making theory. Based on our research into process design and process management, the book offers practitioners an appreciation of the complexity of decision making, suggesting to them how to act, given this complexity. It also presents those with a more theoretical interest in problems, decision making and management with significant views on concepts such as networks, unstructured problems and decision making, thus enriching existing theories.

Hans de Bruijn
Ernst ten Heuvelhof
Roel in 't Veld

CONTENT

1. PROLOGUE

1.1 INTRODUCTION

In recent years, the world of management has given considerable attention to the process aspects of management and change. Many changers define their own role as that of a process manager, process supervisor or facilitator. A frequently heard complaint about change processes is that too little attention is paid to the process (and too much – complainants add – to the substance).

This attention for process management tends to be based on a distinction between substantive and process aspects of changes. Substantive aspects concern the question *what* changes are desirable, process aspects concern the manner *how* these changes can be recognized and implemented. In many cases, the substantive aspects of change are found to be so complex that express and targeted attention for the process aspects is necessary to complete a change process successfully.

This book considers the process aspects of changes. It is based on the idea that substance and process are, indeed, distinguishable but that they are inseparable as well. The design of a process is important for its substance: processes produce substance. The designer of a process who is interested in substantive issues is bound to link the two, both before and during the process.
This book is based on a study of the literature as well as empirical research. This introductory chapter presents an initial description of process management. Chapter 2 positions process management vis-à-vis a number of other management styles.

1.2 THE PROCESS APPROACH TO CHANGE: A TENTATIVE DESCRIPTION

Experiences with change processes tend to be disappointing. A change process is initiated from a substantive design (in the form of a plan, a memorandum, a view, a strategy, a range of objectives, et cetera), little of which is left at the end of the change process. The change achieved differs sharply from the change desired or nothing has changed at all. Eventually, many parties involved tend to be highly dissatisfied with change processes. They usually express this dissatisfaction in substantive terms: 'a colourless compromise', 'a decision without the prospect of proper implementation', 'a planning disaster', 'nothing has changed in comparison with the old situation', et cetera.

What is the cause of such disappointing experiences? We will offer a number of reasons, which we will discuss in detail in chapter 2.

The first reason is that the person initiating and wishing to effect the change tends to function in a *network of dependencies*. He depends on others and can never impose a change unilaterally, since the others can amend, obstruct or even block the intended change, both during the phase in which it was decided to introduce a change and during its implementation.

- The second reason is that the problems are so complex that there is no *unequivocal substantive solution* for them: there is no one right solution. Accordingly, the person who wants a change and produces substantive arguments for it lacks the power to convince the other parties. These parties have other definitions of the problems and solutions and have their own substantive arguments for them. If no unequivocal problem definition and problem solution are possible, further research and analysis tend to be fruitless; they will merely confirm what each party believes to be right or increase uncertainty.

- The third reason is that many changes are structured as projects: accurate problem definitions, clear goals and targets, tight schedules. In a network, however, *a project approach* has *limited significance*, because the parties on which the initiator of a change depends will not automatically accept the latter's problem definition, aims and plans.

- Essentially, the initiator of a change depends on other parties, who fail to be convinced by his substantive arguments, miss much of their own views in the change proposed and will therefore frustrate the project planning. These other parties will not support the problem definition and the solution until they recognize their own views in them. This demands interaction between the parties: they need to consult with each other and negotiate about problem definitions and solutions.

Once the parties realize that they can achieve change only through a process of interaction, they can make rules of the game for such a process. This is called a process design:

Rules of the game that the parties involved will follow to reach a decision.

We would observe here that it is not always easy to reach agreement about such rules of the game. Why should a party commit itself to a number of rules of the game if it does not accept, for example, the problem definition and the problem solutions suggested by the initiator of the change? This is even more so if the party involved knows that the solution cannot be implemented if it simply refuses to cooperate. This party will only be prepared to arrange rules for decision making if such rules leave sufficient room to promote its own interests. Hence, rules of the game will always have to hold out such prospects for the parties involved.

A process approach to change therefore implies that:
- agreements about the decision-making process are made between the parties involved beforehand;
- the emphasis in these prior agreements shifts from the substance of the change to how it will be effected;
- these agreements leave each of the parties involved sufficient room to promote their own interests.

The key difference between a substantive and a process approach to change can be summarized as follows:
A substantive approach to change resembles scientific methods and starts with a substantive design: a substantive problem analysis and a substantive solution. This is followed by a process of introducing the substantive design, which will have succeeded once the substantive design has been realized.
The design and its introduction are based on substantive guidelines (e.g. guidelines for research or for sound engineering); only limited attention is paid to rules of the game.

A process approach to change starts with the parties arranging a number of rules of the game, followed by a decision-making process in line with these rules of the game. This process results in a substantive design.
The design and its introduction are based on rules of the game; only limited attention is paid to substantive guidelines. But in a process, each of the parties involved will, of course, have its own substantive orientation.
This distinction naturally requires considerable differentiation, which this book will make. The above characterization should suffice here.

	Substantive design	Process design
Core element analysis, preceding the design	A substantive analysis of the problems	An analysis of the parties involved, their interests, resources and views.
Core element design	A substantive solution for the problems	A description of a process leading to a solution of the problems
What is the problem?	Is defined in substantive terms where possible. It is therefore known at the start of the decision making regarding a concrete project.	Is known at the start of a process, but may change during the decision-making process.
Key problem regarding acceptance	After the design process: it may be found that the design lacks acceptance	Prior to the design process: acceptance of the process takes time.
Type of decision rules	Many substantive guidelines and a few rules of the game for unforeseeable circumstances.	Many procedural rules and a few substantive guidelines for subjects deserving protection, irrespective of the nature of the project.
Role of the 'change manager'	- *Architect*: makes a substantive design - *Manager*: sees to introducing the substantive design - *Keeper* of a limited number of process preconditions	- *Process architect*: designs the process approach - *Process manager*: manages the decision-making process - *Keeper* of a limited number of substantive preconditions.

Table 1.1 Differences between substantive designs and process designs

A simple illustration may clarify the difference between substantive designs and process designs.

Suppose a new connection has to be built between two areas in a country. A substantive approach will first make a substantive analysis of the underlying problems: what is the volume of the demand for transport, what is the nature of the demand for transport (transport of people, various types of goods transport) and what are the available solutions (road, water or rail, for example). The transport problems and the various solutions are analysed. These analyses

produce a particular outcome, for example in the form of a proposal for the route of a railway line that will be used by twenty trains per hour. Half of these are goods trains, the other half being passenger trains.
In a process approach, the main thing is not the substantive analysis. A process approach maps which parties have an interest in the decision making and what their resources are. These stakeholders are mapped and it is examined what views they hold. A number of process agreements are then made with these stakeholders: rules of the game, indicating how they will reach a decision.
This process of decision making has an agenda: the taking of a decision about a connection between the two areas. This agenda is less directive, however, than in a substantive approach. This is because stakeholders may redefine the problem during the process. They may find that the existing connections will suffice if they are upgraded here and there. The money thus saved may be spent on improving the IT infrastructure in the two areas, for example.
In the substantive approach, the problems start after the analysis has been completed and the proposal for a solution has been made. It may then become apparent that there is considerable resistance to the solution, as a result of which the analysis may have to be started all over again. These problems also occur in the process approach, but they are eventually overcome because the main stakeholders are involved in the decision making.
Changes within a substantive approach require three roles to be filled. The substantive change has to be defined (architect), the introduction of this change has to be managed (manager) and there are of course a limited number of procedural preconditions to be met. The opposite is true in a process approach; the architect designs a number of rules of the game, manages the decision making and has a limited number of substantive preconditions.

The process manager and the process architect

As outlined above, a shift from a substantive approach to change to a process approach means a shift in the role of the 'change manager'. The change manager is primarily a process manager [1]. He sees to the change process, ensuring that the parties involved follow the rules of the game, that there is proper communication between them, that decisions are made in conformity with the rules of the game, et cetera.
The "process architect" also has a role to play. In a substantive approach, a design will be made by substance experts. A process approach requires a process architect, responsible for laying down rules of the game. It is for the process architect to ensure that the process design appeals to the parties involved: it should offer them sufficient opportunity of promoting their own interests.

1.3 STRUCTURE OF THE BOOK

This book consists of three parts. Part I contains a number of introductory ideas about process management. Chapter 2 contrasts process management with a number of other management styles.
Presenting the main arguments for a process approach, it also deals with the main risks.
Part II deals with process design. Chapter 3 contains a number of principles that may be used in designing a process. Chapter 4 focuses on the actual design of a process: what are the key tasks of a process architect leading to a process design?
Once a process design is available, the process will need managing. Part III highlights such process management. Chapter 5 sets out how to guarantee the *openness* of a process. This openness concerns the actors to be involved and the subjects to be placed on the process agenda. Chapter 6 discusses the *protection of the actors' core values*: how to prevent a process from harming an actor's essential values, which is undesirable because it may very well cause this actor to frustrate the process. Chapter 7 sets out how to give a process *sufficient speed.* After all, processes can lead to extremely slow decision making. Chapter 8 discusses *content* in a process. How can a process be managed so as to produce substantively rich results? How can we prevent the result being substantively poor for the sake of agreement? A brief epilogue concludes the book.

Notes

1. The notion of change management as process management can be found in several places in management literature: in the literature about change management in professional organizations (for example, D.U. Buchanan and A. Huczyncky, Organizational Behavior: An Introductory Text, Hertfordshire, 1997), about networks and network organizations (e.g., Hans de Bruijn and Ernst ten Heuvelhof, Networks and Decision Making, Utrecht, 2000), about consensus building and mediating (e.g. J.E. Innes, Planning Through Consensus Building: A New View of the Comprehensive Planning Ideal, in: Journal of American Policy Analysis (1996), vol. 62, No. 4, pp. 460-472; L. Susskind, S. McKearnan and J. Thomas-Larmer (ed.), The Consensus Building Handbook, Thousand Oaks, 1999), about stakeholder management (C. Eden and F. Ackermann, Making Strategy, London, 1998, p. 4).

PART I
INTRODUCTION TO PROCESS DESIGN AND PROCESS MANAGEMENT

2. POSITIONING THE PROCESS APPROACH

2.1 INTRODUCTION

This chapter contrasts the process approach with four management styles.[2] The process approach is contrasted with substantive steering in section 2.2, with a management style marked by command and control in section 2.3, with project management in section 2.4 and with single-value management in section 2.5.
In section 2.6, this positioning results in an overview of the main arguments for process management. We will also set out what variants are possible in how parties support the results of a process (section 2.7).
We will then discuss a number of risks posed by process management. The first risk is that process management is used as a management tool. The second risk is that process management completely eclipses the contrasting management styles. This would leave no room at all for command and control and substance, for example. This may lead to sluggish decision making and affect the quality of decision making: involving many actors causes the result of a process to degenerate into a colourless compromise (section 2.8).

2.2 PROCESS MANAGEMENT VERSUS SUBSTANCE

First, process management is the opposite of a substantive approach to decision making. The underlying idea is that some problems are 'unstructured'.
We define unstructured problems as problems for which no unequivocal and/or authoritative solution is available. The reason is twofold:

- no *information* is available that can be measured objectively, and
- there is no consensus about the *relative weight of the criteria* used in problem solving. [3]

The first problem arises because discussion is possible about the data, methods and system boundaries (i.e. the delineation) used in the problem definition and the choice of the solution. The second problem arises when, to reach a solution, a trade-off has to be made between several criteria and discussion is possible about the weight attached to each of these criteria.

> The following example illustrates how an unstructured problem was solved with the help of a process design. Urged by politicians, businesses want to decrease the environmental impact of packages for consumer products. To do so, they have to examine the environmental harm caused by the various packages. Thus, a dairy factory wants to know which of the following is the least harmful form of packing: the glass bottle, the carton box, the polyethylene bag, or the polycarbonate bottle.

The environmental impact can be measured by comparing the various packages in *terms of environmental compartments*, including toxicity, energy consumption, photochemical smog, emissions and waste.

Although the environmental impact might seem easy to determine, actually it is not. The first problem is that objective information is hardly available. To measure a package's energy consumption, one must establish how much energy it takes to produce a package, to transport it to the consumer, to return it, if desired, et cetera. Measuring these objectively is hardly possible, if at all. *Data* is lacking, some of it is obsolete or is over- or under-aggregated. *Methods* to calculate a package's environmental impact are disputed and can never be fully objective. Furthermore, the problems' *system boundaries* are debatable: when defining the environmental impact of the transport of raw materials, is the environmental impact of the production of the means of transport (ship, lorry) also taken into account, for example?
Once information has been gathered about the environmental impact for each of the compartments, a package must be chosen. Which package is the least harmful to the environment? One that scores well on energy, less well on emissions and neutral on waste? Or a package that scores poorly on energy,
neutral on waste and well on emissions? This is the second problem: there are no unequivocal criteria to compare the individual environmental compartments.
It will come as no surprise, then, that different parties hold different views about the environmental impact of one and the same package. Nor will it come as a surprise that each of the parties has an independent study available, confirming what each believes to be right. [4]

The unstructured nature of problems also stems from the fact that some problems cannot be solved in isolation, but are *interrelated* with others. Consequently, solving problems demands a trade-off, not only within a defined problem (in the above example: the environmental impact of packages), but also between different kinds of problems (such as the environment, the economy, safety). Of course, making trade-offs is never a neutral, objective affair.

In the above example: the choice of a package is based not only on environmental concerns, but also on considerations of cost and safety, for instance. Moreover, every environmental improvement may raise the question of cost-effectiveness: what environmental improvement will an environmental investment worth X lead to? And might the same investment lead to greater environmental improvement elsewhere? Again, an objective trade-off between cost, safety and environment is hardly possible, if at all. [5]

The third characteristic of the substance of problems requiring a process design is that they are *dynamic*. The problem changes in the course of time. The logical consequence is that the answer to the question whether something is a solution for a problem also changes in the course of time.

> Assume that the environmental impact of a package in the above example has been defined for the individual environmental compartments (energy consumption, emissions, waste, et cetera). Also assume that there is consensus between companies, government and societal organizations about the criteria used (1) in the trade-offs between the various environmental compartments, and (2) in the trade-off between the environment, the economy and safety.
> Several developments may now take place that call the defined environmental profile into question. Technological innovations may be introduced. For example, it appears technically possible to generate more energy from burning package waste than before. As a result, a package scoring poorly on energy may suddenly be found to score much better.
> Furthermore, producers of packages that scored badly may endeavour to improve the environmental profile of their package by introducing energy-extensive production processes, for example.
> In the above examples, the information used earlier has become obsolete due to technological innovations. Criteria may also change. Suppose a country has a serious shortage of waste incinerators. This will cause the reduction of waste to become an important standard for appraising packages. This opinion on the relative weight of criteria may change. Once the capacity shortage is alleviated, the relative importance of energy may rise again.
> This change may even alter the nature of the problem. The intelligent problem solver will find that the problem is not so much what the most environmentally friendly package is, but how to keep the process of the ongoing improvement of packages going. If the problem is the most environmentally friendly package for milk, one of the milk packages will be chosen. If the problem is how to keep the process of the ongoing improvement of packages going, the solution might be that all four milk packages have the right of existence and that they have to keep the environmental impact as low as possible in mutual competition.

Why are problems *dynamic*? The above reasoning already partially answers the question. New *information* may be released (about the availability of technological innovations, for example) and opinions *on the relative weight of the various criteria* may change (for example: energy consumption is less important than water consumption).
This dynamic means that attention shifts from finding a correct problem definition and solution to an *ongoing process* of formulating and solving problems. Any solution found today may be obsolete tomorrow.

Those who choose a substantive strategy of change regardless of the unstructured nature of problems will only create conflicts. Instead, an initiator should accept that different parties use different definitions of reality and may have good arguments for doing so. A solution will never be capable of objective assessment, but it may be *authoritative*: it is then accepted by the parties. This requires involvement of the parties in the problem-formulating and problem-solving process. Problem definitions and solutions can then be authoritative, because they are *negotiated knowledge*: knowledge resulting from a process to which the parties have contributed their own information and values.

2.3 PROCESS MANAGEMENT VERSUS COMMAND AND CONTROL

Decision-making processes tend to take place in a network. Table 2. 1 contains the main characteristics of networks, compared with the characteristics of a hierarchy.[6]

Hierarchy	Network
Dependence on superior	Interdependence
Uniformity	Pluriformity
Openness	Closedness
Stability, predictability	Dynamic, unpredictability

Table 2. 1 Hierarchical and horizontal management

When a decision-making process has to take place in a network, this always implies that several actors are involved in the decision making. They have different interests and are interdependent. No single actor can itself fully realize its own goals (interdependence).

However, there are many differences between actors (pluriformity), hampering cooperation and concerted decision-making. In particular situations, particular actors may not be interested at all in cooperating with others (closedness), which hampers decision making even more. Finally, the number of actors involved may change in the course of the decision-making process (dynamic): actors join and leave.

In a network, hierarchical management stands little chance of success. A manager who wants to realize a project through command and control may seem energetic and powerful, but usually lacks the knowledge and power to realize his own views and will consequently meet with considerable resistance in the network. Other parties are able to block, delay or change his project. This may make hierarchical management highly counterproductive: the manager may seem energetic, but merely provokes resistance. The more hierarchy, the stronger the resistance in a network. Goals remain unachieved and plans unrealized.

The opposite of command and control is a process approach. As soon as a manager has to function in a network, he cannot rely on hierarchical managing mechanisms. After all, he depends on other parties, which will not automatically support him. A manager who recognizes this, will not take decisions unilaterally, but reach a decision in a process of consultation and negotiation with other parties. Such a process reflects the mutual dependencies in a network. The literature refers to it as 'interest-based' decision making.[7] It is worth noting here that an interest is seen as a legitimate perspective of reality (not as an ordinary and one-sided perspective).

> Thus, the packaging problems might seem a matter for business. But government and societal organizations like the environmental movement want to address these problems as well. Companies partly depend on government and societal organizations. If, for example, the environmental movement feels that industry has chosen the wrong packages, it may seriously disrupt corporate marketing. It may seek support from a government, which may use its legal instruments. Packages are of strategic importance for the sale of consumer products.
> Being under constant fire from societal organizations will harm a company's image and may even lead to a fall in demand. The conclusion is that companies partly depend on the support of government and societal organizations when choosing a package.[8]

This does not mean, however, that in networks there is no role for command and control. This management style can play a role in the process approach (see chapter 7), but the dominant idea is that networks force those involved to adopt a process approach to decision making.
It is important to note here that parties in networks tend to display *strategic behaviour*: They play games to advance their interests.[9] For example, they may make public the information they have available piecemeal, or make it public at a time that suits them best. They may be deliberately vague about their own views, values and criteria to keep a free hand in the decision-making process. They may form coalitions with other parties: for example, mutual arrangements are made, allowing parties to act *en bloc* and reinforcing the power of these cooperating parties.
Evidently, such strategic behaviour obfuscates decision making even further. A manager who fails to recognize this and provides command-and-control-type steering will again achieve the opposite of what is envisaged: in many cases, he

will merely provide incentives for strategic behaviour. Process management might seem to be a wiser strategy.

> We give an example of this from our study into the rise and fall of the Sport 7 TV channel. The Royal Dutch Football Association (KNVB) launches a sports channel (Sport 7). For the success of this channel it heavily depends on other parties, including the paid-football organizations, cable companies, advertisers, competitors (NOS, Studio Sport) and several departments/ministeries. KNVB sets a deadline for the launch of the sports channel: Sport 7 is to be fully operational at the start of the football competition, despite resistance to the channel. This attitude might seem energetic, but is actually an incentive for resistance by the opponents of Sport 7. Blocking the sports channel is highly attractive for them: if it is not operational at the start of the competition, KNVB will have a problem, while blocking it will reinforce the opponents' position.[10]

Incidentally, it should be observed here that it is not always a problem if particular parties refuse to cooperate in a change process. After all, such a process is initiated to change the existing situation and particular parties may see their position deteriorate in the course of it.

> An example of this is a reorganization in a company. Such a reorganization tends to require involving the company's main stakeholders (such as the directors of the various departments). They know the primary process better than the top of the organization. Thus, managing reorganizations often has the characteristics of process management. But there are also limits to involving stakeholders, since one or more departments may be sacrificed in a reorganization. Hence, involving stakeholders does not imply that their interests can never be harmed (also see chapter 6).

The situation is altogether different, however, where these parties hold a position of power, making their cooperation essential for a successful implementation of the change process. In that case, agreements will have to be made with such parties so that they will commit themselves to the change.

2.4 PROCESS MANAGEMENT VERSUS PROJECT MANAGEMENT

Thirdly, a process approach can also be positioned vis-à-vis a project approach. In a project approach it is assumed that problems and solutions are reasonably stable within certain limits.
This allows the use of project management techniques: clear goals and targets, a time schedule, a clear framework and a prefixed end product, all of which lead to linear and structured decision making. Naturally, such an approach will only work in a static world. When an activity is dynamic rather than static, a project

approach is impossible and a process approach is desirable.[11] This dynamic may have both external and internal causes.

- External dynamic: an activity starts as a project, but develops into a process because external parties interfere with the project, contributing their own problem definitions and solutions. This is the familiar course of events in many infrastructural projects. What starts as a project (e.g. building a stretch of railway line) ends as a process in which all sorts of parties interfere with the railway line and which has its own dynamic. After some time, the parties may have given up discussing the railway line and may now be discussing completely different subjects.
- Internal dynamic: an activity starts as a project, but develops into a process because, during the project, the project owner learns that the problem is more comprehensive or more complex than he thought. A nice example is that of a houseowner, who one day decides to hang a painting somewhere else on the wall, finds that the colour of the wall behind the painting has faded, redecorates the whole wall, which affects other parts of the interior and eventually leads to the whole house being refurbished. He then finds out that this renovation urge has to do with the phase of life he is going through and finally ends up with a psychologist. What starts as an simple project ends in a complex process involving many parties (i.e. other members of the household, builders, neighbours, psychologist).

There will be dynamic particularly when decisions have to be taken in a network. The various parties hold different views about how a problem and a solution should be defined. As a result, the decision making will always be capricious and unstructured.

Table 2.2 contains the differences between decision making in a hierarchy and in a network.[12]

Hierarchy	Network
Regular	Irregular
Phases	Rounds
Actors are stable, behave loyally and are involved in formulating the problem and choosing a solution.	Actors join and leave, behave strategically; there are often winners and losers when the problem is formulated.
Starting point and end clear	No isolated starting point and end
Problem → solution	Solution → problem

Table 2.2 Decision making in a hierarchy and a network

A linear and structured decision-making process, from noticing a problem, through a number of phases, to a solution, fits into a hierarchy. The decision-making process is initiated by the actor who is superior in the hierarchy. The other actors participate in this decision-making process, behave cooperatively, partly due to their hierarchical subordination to the actor formulating the problem. Much of the decision making is a matter of project management.
There is no project-type and phased development in a network, however. Although the initiating party may formulate problems, they do not automatically become the subject of decision making, nor are they always solved. The reason is a simple one: the other parties see no sufficient reason to place the problem on the agenda or lose interest in the problem during the decision-making process.
During the decision-making process, these parties may also find that the solution that seems available will harm their interests. They may try to block any further decision making. In other words, the parties participating in this managerial process behave strategically, partly due to the absence of a hierarchical subordination.

> An example here is the decision making about large infrastructural projects.
> Assume that the municipality of Rotterdam and the Rotterdam Port Authority want to create extra industrial estates in the port of Rotterdam through land reclamation, allowing the building of a second Maasvlakte area. According to the Rotterdam authorities, the underlying problem is the shortage of industrial estates in the port area. It is not so difficult to imagine that a project-type approach stands little chance of success. For land reclamation, Rotterdam depends on the support of other parties. Some parties will refuse to accept that there is a lack of space, other parties may recognize the problem but envisage completely different solutions, i. e. better utilization of the existing space rather than expansion, expanding the port of Vlissingen (Flushing). The stronger the city of Rotterdam's project-type steering and the stronger therefore the emphasis on its problem definition (lack of space) and problem solution (land reclamation), the further the chance of realizing it may diminish. For parties that cannot identify with it, there is not a single incentive to support Rotterdam. They will put up either passive or active resistance, causing not linear, but capricious and unstructured decision making.

This is one of the reasons why a regular and linear decision-making process may be replaced by a process taking place in rounds.[13] In a round, actors reach a decision during a fight, or rather try to prevent doing so. A round will finish at some time and produce a provisional result, including winners and losers. This might seem to conclude the decision making, but a new round may suddenly present itself.
The behaviour of the parties involved explains this capricious course taken by the decision making. We will mention a number of examples of behavioural patterns

that are absolutely rational in a network and lead to the capriciousness referred to above. For each of them, we will explain the developments on the basis of the example about land reclamation.

Dynamic

New developments may take place, as a result of which problems or solutions can be redefined. Parties that regard these new developments as a chance of influencing the decision in their favour will introduce them into the decision-making process.

> In the Rotterdam area, the 'recreation' item may feature high on the agenda, as the area lacks recreational facilities. This may lead to the problem being redefined: it is no longer just the lack of space, but also the area's quality of life. This presents possibilities. Support for the Maasvlakte area may be swapped for Rotterdam's support for expanding recreational facilities. Or part of the reclaimed land may be designated as a new recreation area. In either case, the chance of realizing the Maasvlakte area will increase. Evidently, a project-type approach will hardly be effective here. A project-type approach fails to avail itself of the opportunity presented by changing the problem definition.

Compensation of losers and couplings

All decision making will produce winners and losers. In a new round, losers can try to compensate their loss and so try to frustrate decisions that have been taken. But the winners may also take compensatory action: they may compensate the losers of decision-making process X in a following decision-making process Y. This creates a coupling between X and Y, which will seriously hamper a project-type approach to subject Y.

> The submunicipality of Hook of Holland, a seaside resort on the North Sea, is one of the potential losers in the decision-making about the Maasvlakte area, since the land-reclamation development might be visible from this submunicipality. It is likely to receive compensation for this loss, for example by means of an expansion of the number of 'bad-weather facilities'. This will make Hook of Holland more attractive for holidaymakers and daytrippers. Suppose one of the forms of compensation is the building of a swimming pool. This couples the decision making about land reclamation to that about a swimming pool in Hook of Holland. This coupling to Hook of Holland's problems may help to realize the land-reclamation project, but also shows that a linear, project-type approach is hardly fruitful.

Solution seeks problem

When the parties have chosen a solution, they tend to start looking for attendant problems. Proving that their own solution solves another party's problem may gain them the support of this other party.

> Building an airstrip in the Maasvlakte area may take some of the pressure of Amsterdam Airport Schiphol, which has major capacity problems. The Maasvlakte solution is then coupled to the Schiphol issue. This may be an attractive strategy for Rotterdam because it might thus win the support of particular actors. Again, this is an effective strategy, which cannot be understood from a project-managerial perspective, however, because it makes decision making more capricious. It involves not only lack of space and land reclamation, but also the airport's infrastructure.

Blocking power in the tail of the decision-making process

Actors have been known to become active only in the tail of a decision-making process: they see what solutions are likely to be chosen, have difficulty in accepting them and try to block or adjust any further decision making. To use the terminology of the rounds model: they present themselves in the final rounds and only then do they participate in the decision making.

> The province of Groningen may feel at some stage that national resources are finding their way to Rotterdam to fund the land reclamation. Groningen may try to influence the government's decision making by arguing that state funding of a port infrastructure calls for an integrated trade-off between the Dutch ports, including Eems port in Groningen. This view may be unlikely to meet with acceptance, but it is characteristic of decision making in networks. In the tail of the process, parties present themselves that may be found to have an unexpected amount of blocking power.

Strategic behaviour

Finally, we should point out that actors may adopt strategic behaviour. Those who know that decisions are taken in rounds can adapt their strategic behaviour accordingly. Actors may, for example, decide to adopt a reserved attitude in round X and accept a loss in order to have a stronger position in round X+1. The reverse is possible as well: heavy resistance in round X, so that compensation can follow in round X+1.

> The port of Vlissingen and the province of Zealand take part in the land reclamation game. Resistance to Rotterdam's expansion is unlikely to be effective, but may nevertheless be interesting for strategic reasons. The stronger Zealand and

Vlissingen's resistance, the better the chance of compensation in the event of future loss.

Each of these behavioural patterns shows that project management stands little chance of success. Project management may even be counter-productive because it fails to avail itself of the opportunities that capricious and unstructured decision making affords. The opposite of project management is process management: an initiator realizes that he depends on other parties, invites these parties to join in a negotiating process in which these parties couple problems and solutions. This process will take a number of unexpected turnings as a result of the behaviour of these parties, but the damage done by these turnings may be reduced or these turnings may even be utilized so long as the parties to a process negotiate.

2.5 PROCESS MANAGEMENT VERSUS SINGLE-VALUE DECISION MAKING

Robert Quinn introduced the concept of the 'management of competing values'.[14] His theorem is that, for an organization to be effective, it often has to respect competing values in its management. It should not choose between central management or decentralized autonomy, for example, or between hands-on and hands-off management, but strive to unite these and similar competing values. An organization that is able to realize strong central management as well as decentralized autonomy is more effective than an organization that opts for central management only or decentralized autonomy only.

This *management of competing values* can be realized in a number of ways.

- *Alternating*
  The various values alternate in the course of time. Periods in which central management dominates alternate with periods in which decentralized autonomy dominates.
- *Compensating*
  Management styles are used simultaneously and compensate each other's shortcomings. Where decentralized autonomy fails, central management is chosen; if central management is not effective, decentralized autonomy is granted.
- *Competing*
  An inconsistency in the design of an organization is tolerated deliberately. There is room for both central management and decentralized autonomy and it is not stated explicitly when one value dominates and when the other dominates. The tension this creates in an organization is perceived as positive: it will make the proponents of decentralized autonomy alert to the

shortcomings of autonomy and the proponents of central management to the shortcomings of decentralized autonomy. The tension between the two management styles is thus a fertile breeding ground for a well-performing organization.

Empirical research confirms the picture painted by Quinn: an organization with competing management styles has a fruitful tension; single values lead to a one-sided management style, which in many cases is unable to deal with its own shortcomings or with changed circumstances.[15]
The reason is that many organizations are *hybrid*: the organization is based on mixed and – in many cases – competing values.[16] Examples of these are public values versus private ones and hierarchical values versus contractual ones.

We shall give an example of this range of values.

- A university generates knowledge and makes it available to society. It receives public funding and making its knowledge freely available is therefore an important public value. The university also does a lot of contract research: research commissioned and funded by third parties. Entirely different values may apply here: the knowledge generated is made available to the commissioning party only.
- A ministry is under political, hierarchical control. The minister is the competent and thus responsible authority. For a civil servant, the corresponding values apply, such as official loyalty. The same civil servant will often have to function in negotiations with societal players. Different values may apply here, such as good faith between negotiating partners.

The key motive for hybridity is always synergy: mixing the competing values generates an advantage that cannot be achieved in a one-sided value orientation.[17] In 't Veld argues that many forms of hybridity are found particularly in the innovating compartments of society, whereas innovation stagnates in sectors where hybridity is counteracted and single-value systems dominate, such as education, health care. [18]

Synergy in the above examples.

1. Contract research and publicly funded research can reinforce each other. Contract research offers the possibility of keeping in touch with societal developments (some of which may take place very fast) and testing developed concepts and theories. Publicly funded research allows reflection on the findings of contract research, broadening and embedding them theoretically.

> 2 Negotiating management is an important and – in many cases – necessary complement to hierarchical forms of management. Negotiations offer other possibilities of influencing behaviour, make a government more susceptible to societal resistance and can therefore improve the quality of government intervention. Hierarchical interventions may be highly counterproductive (also see section 2.3). Hierarchical interventions are nevertheless necessary, since negotiating management can also cause results to be delayed or a government to become the prisoner of its negotiating partners. It is the combination of hierarchical and contractual values that is attractive.[19]

Consequently, the management of competing values improves a hybrid organization's performance. The opposite of a single-value orientation is a multiple-value orientation. Hybridity carries two important risks, however. The first risk is that the balance between the values is disturbed and one of the values will dominate; value X forces out value Y. The underlying idea is that the value that is simplest to realize will force out the other value. The second risk is that the *management of competing values* does not create synergy, but has perverse effects. The underlying idea then is that the two values are incompatible in principle.

> A university that can acquire relatively simple contract research runs the risk that this research will start to dominate other forms of research. In a political culture in which separation is a dominant value, it is difficult to maintain the value of integration.
> Examples of perverse effects of hybridity are distortion of competition between the university and private companies; alternating hierarchical and contractual management at random: a minister using hierarchical management or management through contracts as he sees fit and thus losing the confidence of societal parties.[20]

When it is recognized on the one hand that hybridity and the ensuing competing values are meaningful, while on the other hand it carries important risks, two strategies are possible in response to this. Process management is an important component of both strategies.

The first strategy is structural intervention in hybrid organizations. This first reaction was formulated by Cohen, among others, and focuses on organizations carrying out public and commercial tasks. Mutatis mutandis, it also applies to other forms of hybridity.[21] The reasoning goes as follows. If an organization performs a public task and charges no price for it or is a monopolist, there is no market discipline. If the same organization also operates on one or more standard markets, this entails the risk of cross subsidies. This is a perverse effect of hybridity, because it leads to unfair competition. According to Cohen, adequate checks whether this perverse effect occurs in the hybrid organization are impossible. That is why the principle of structural separation between the values is inevitable: split up hybrid organizations and make sure each value is propagated by a different organization.

The second strategy is based on the idea that the advantages of hybridity will be lost in a structural separation of values. Structural integration is necessary. In 't Veld gives the example of an operation in a university hospital.[22] A team carries out the operation in the operating theatre. This operation is embedded in a large number of productive processes:

- Naturally, the operation is a form of health care;
- for the operating professor, the operation forms part of his clinical research;
- for some nurses, the operation forms part of their training to become operating-theatre nurses;
- for the student surgeon and the student anaesthetist, the operation also forms part of their training;
- for the surgeon and the anaesthetist, the operation is necessary to maintain their professional expertise and skills.

If, for some reason, structural separation was mandatory here, each of the productive processes would become many times more expensive and the same quality could not be achieved. In short, structural separation is not an option.
The strategy then is to preserve hybridity in the organization and find a solution for the perverse effects that may occur. In an organization, a culture should be developed in which hybridity is recognized and appreciated, but in which the perverse effects of it are also recognized and positive combinations of the values are sought where possible. Hybridity in organizations also has its consequences for the form of these organizations' external relations. They should be designed so as to allow these external relations to contribute to the prevention of perverse effects of hybridity.

> We will give a number of examples. A university may base its relations with the minister responsible for public funding on contracts. It then enters into obligations to the minister which should adequately protect public values. This creates uniformity in external relations: in addition to private contracts there are public contracts, which may improve the balance between the two values. If a minister constantly tends to ignore the results of negotiations by hierarchical interventions, there may be a rule of the game dictating that an agreement between the minister and societal parties should be submitted to parliament for approval on the assumption that this makes it less easy to set such an agreement aside.

We are not concerned here with the question which strategy is preferable under what conditions, but with the fact that both strategies demand forms of process management. A strategy of structural separation carries the risk that the advantages of hybridity are lost. If a structural separation of tasks is chosen, the next step necessary is that the separated organizations invest in a process of cooperation. In other words, structural separation and disconnection are accompanied with an integration and interaction process. This interaction process implies that a number of rules of the game for cooperation are formulated to prevent losing the synergetic advantages of hybridity.

> Examples can be found in many utility sectors where liberalization and privatization are taking place. Many national railway companies have now been subjected to competition. In a monopoly, these railway companies filled at least three functions: they managed the railway network, provided services and regulated rail traffic. In an open market, this is a problematic combination of tasks. On the one hand, having these three functions in one hand is a value: it guarantees integration
> On the other hand, it hampers the creation of competition. The decision in this case was to terminate structural interrelation. Service provision has been separated from network management and traffic control.
> Naturally, proper rules of the game between these three organizations become necessary then. Many of the problems with service provision that Dutch Rail (NS) is facing at the start of the 21st century are due to flawed rules of the game between NS as a service provider and traffic control, organized separately. Rules of the game between the two organizations are insufficiently specified and internalized.

The risk of a strategy of structural integration is to create the forcing-out phenomena and/or perverting effects mentioned above. If structural integration is chosen, it is necessary to shape accountability to third parties (superiors, principal, parliament, customers, etc.) so that it will be clear to these third parties whether this forcing-out or perversion has occurred. They can form an opinion about it and then intervene. So this, too, requires a form of process management: rules of the game for the process in which account is given and any subsequent interventions have to be specified, so as to minimize the risks of a structural integration.

> A university that opts for mixing its public and private functions can design an accountability system that minimizes these risks. Market-related rates are used, there is an integrated time-recording system, it has to be indicated for each project how it can be coupled to publicly funded research, an aggregated report on the projects has to be made every two years, et cetera.

In essence, the reasoning here is as follows. A single-value orientation does not match the hybrid nature of many organizations. A multiple-value orientation is desirable, *management of competing values*. The two strategies to realize this

multiple orientation are structural separation or structural integration. Both strategies require complementary process arrangements. A structural solution requires process management if it is to reach its aim.
In conclusion, we would point out here that this positioning of process management vis-à-vis the four competing management styles does not imply that these styles have no value. On the contrary, in process management, too, command and control, substance, project management and single-value decision making may be necessary. We will revert to this in section 2.8 and in part II of this book.

2.6 THE MAIN ARGUMENTS FOR PROCESS MANAGEMENT

The above positioning of process management leads to a number of arguments for process management.

Support

Decision making involves many parties, many of which have blocking power: in some cases they may cause decision making to stagnate for a long time. For these parties to lend their support, they should be involved in the problem-formulating and problem-solving process.

> Thus, the Directorate-General for Public Works and Water Management uses the *open-plan process* in the decision making about projects. Three frequently asked questions about this open-plan process are:
>
> - *"Is the environment allowed to influence the results?* Yes. The environment is not only allowed to participate in the thinking about the results (the solution), but also about the problem definition. So the open-plan process is neither 'propaganda' nor a manipulative method.
>
> - *Does negotiating form part of the open-plan process*? Yes. The open-plan process aims to couple goals and interests from the environment to our own goals and interests. This automatically places Public Works and Water Management in a negotiating position. Depending on the type of project, the environmental characteristics and the substance, Public Works and Water Management can adopt a more or less open attitude in the negotiating process. It makes a lot of difference here whether the problem definition of the project is at stake, or the direction for a solution or the decision about it.
>
> - *Is it necessary to reach consensus about the outcome of the process*? Basically, we try to find a solution that everybody agrees with. Unfortunately, reaching consensus is a Utopian dream in most cases. After all, conflicting interests are

> difficult to combine. However, in many cases, a result can be reached that satisfies all parties. If consensus remains out of reach, we will try to find maximum support for the outcome. This is how we try to guarantee the most realistic decision making possible." [23]

It should be pointed out that the argument of support plays a part not only in the public sector. In many large organizations, the management is highly dependent on the professionals who perform the primary process, and for that reason resorts to forms of 'open decision making'. Shell, for example, has introduced the 'Triple D model'. DDD stands for Dialogue, Decide, Deliver and is the opposite of a DAD approach: Decide, Announce, Defend.[24]

Reducing substantive uncertainty

Having all the relevant information available is vital to solving unstructured problems. In many cases, the parties involved have other information, essential for the adequate solution of a problem. Confronting the different sources of information with each other may improve the quality of the information used.
For such a confrontation to take place, the relevant parties should be involved in the problem solution.

> An example of this is the building of a space shuttle: a highly complex project, seen from a technical point of view. A major problem in projects of this kind is *unruly technology* (meaning: technical uncertainty). The building of all sorts of components requires the most advanced technical knowledge, which has – in many cases – been insufficiently developed. The designs used still contain uncertainties, requiring further research or an experiment. There are only limited possibilities to test the technical options chosen. The results of such tests tend to be open to more than one interpretation. In short, no 'hard' and objective knowledge is available to solve the technical problems.
> As a result, the management is faced with many uncertainties and incomplete information during the building of the space shuttle. Choices nevertheless have to be made. Whether the technical choice carries too grave a risk is never certain.
> In such a situation, an organization should strive to minimize the risk of these choices. The jargon contains terms like *debugging* and *closure*. In the NASA organization, this is put into practice by a form of process management. Every risky decision is subjected to a formal *check and double-check* procedure: once the technical specialists have made a choice, it is subjected in the organization to a process of counterchecks, in which the choice is screened. These *review procedures* have been formalized. Any uncertainties that remain after such procedures are formulated as accurately as possible and submitted to the management. The management then has to decide whether a technical risk will be taken, money and time will be allocated for further research within the existing design, or whether a

new technical design will have to be made. This makes process management an important tool for management to uncover technical imperfections. [25]

Enriching problem definitions and solutions

Different parties may have entirely different perceptions of and (normative) views about problems and solutions. Confronting these different views may have an enriching effect.[26] For such a confrontation to take place, the relevant parties have to be involved in solving the problem.
The first argument mentioned for the process approach is gaining support. For some people, support has no positive connotation: it is a necessary evil that will affect the quality of the decision. They see it as good, substantive ideas having to be watered down in order to gain support. The argument of enrichment contests this notion. Support and substantive enrichment of ideas can go hand in hand in this argumentation.

The board or the management of a professional organization wishing to develop a strategy for its organization (i.e. a strategy that influences the organization) will be unable to do so without involving the professionals. Such involvement will raise support for a strategy and tends to enrich such a strategy. The professionals in the organization know the opportunities and threats in the organization's environment better than anyone else and also know what power the organization has to respond to them.

The argument of enrichment not only concerns the product of a process (in the above example: a strategy). Enrichment may also concern the knowledge and values of an individual party. A party may itself develop a richer view thanks to the process. It may also gain a better understanding and a different valuation of the views of the other parties thanks to the process. Both forms of enrichment may contribute to gaining support.

Incorporating dynamic

As we observed earlier, dynamic may cause the chosen problem definitions and solutions to become obsolete fairly soon. This affords unwilling parties the opportunity to distance themselves from a chosen solution, invoking new information, new solutions available, et cetera.
In most cases, the real art is to prevent such new insights and information being available outside the problem-solving process rather than within this process. The only way to achieve this is to involve all relevant parties in the solution-seeking process, since they are the carriers of new insights and information.

Business and societal organizations consult with each other about the environmental impact of packages. It may be important for particular companies or research

institutes to join this process, because in many cases they are the carriers of innovations.
If they do not participate in the process, they will publicize their new insights outside the process at some stage. This can lead to a situation in which parties within the process are consulting with each other about particular options, while parties outside the process have known from quite some time that innovations are forthcoming, making this discussion outmoded. Involving these innovative parties implies that the dynamic can be discussed in the process.
In many cases, this is the reason why parties are willing to consult with highly critical opponents. They prefer this criticism to be levelled in the context of a process rather than being made public outside the process. Compare the process at NASA outlined above: criticism is organized, because it had better be discussed in the context of the process than allowed to circulate in the organization (or worse: outside it). [27]

When a process leads to such incorporation of dynamic, the parties can learn something. This is because they are constantly faced with others and new views, which can lead them to reflect on their own views.[28]

Transparency in decision making
Decision-making processes tend to be extraordinarily complex: many parties, many processes, many subjects. A process design offers a certain amount of transparency. The parties involved can find out at any time where they have got to in the decision making process, what the nature of a decision is, et cetera.

De-politicizing decision making
Change processes tend to provoke heavy resistance. Excessive substantive steering at the start of a change process can boost resistance. A process approach to change can reduce this resistance, since it does not specify the substance of the change, but only what the process towards a possible change will be.[29]

2.7 THE RESULTS OF A PROCESS: CONSENSUS, COMMITMENT OR TOLERANCE

Naturally, the result of a process partly depends on the aims of the process. It may take on all sorts of forms: a number of decisions, a number of actions, a physical product, a system, a strategic plan, etc.
There may be consensus about the results of the process: the parties involved fully agree with each other. The process has been a fair one and the substantive result can count on everyone's approval.
In many cases the conflict of interests is such that consensus cannot reasonably be expected. A process can indeed lead to a commitment being made to a

particular result. A party that makes a commitment declares that, although it commits itself to the result, it does not do so on substantive grounds. A party may issue such a commitment because, for example, it has learned during the process that it has no alternative. Non-commitment would do more harm than commitment.
Commitment implies that a party is willing to contribute to the implementation of decisions. Another step would be for a party to declare that it will not issue any commitment (and so will not contribute to the implementation of a decision), but that it will tolerate the results. This implies that it will not block or hamper the implementation of a decision.

2.8 THE RISKS OF PROCESS THINKING

What are the risks of a process approach?[30]

- The first risk is that an initiator will stick to the classic management style (substance, command and control, etc.) and make process management subservient to it. The process is then merely an instrument to support a classic form of change management. It is thus cast in the mould of the classic management styles.
- The second risk is that a 'swing of the pendulum' is made on each of the three spectra (i.e. process versus content, process versus command and control, process versus project). Substantive management, command and control, etc. no longer play any role; management is completely process-oriented.

2.8.1 Process management cast in the mould of classic management styles

If process management is cast in this mould, the risk is that it will pervert. This perversion can assume two forms.

Explain rather than interact
This means that process management is seen as an instrument to communicate an already taken substantive decision as effectively as possible. Process management is not a means to consult with other parties about prospective decisions, but particularly serves to explain prospective decisions properly. A good and well-considered substantive decision already exists, which, however, encounters resistance yet. To overcome this resistance, it is wise to use a number of process management techniques.

A concrete manifestation of this in organizations is that process management tends to get into the hands of the staff unit of communication. This is because process management has been narrowed down to proper communication, giving a proper explanation afterwards of the wise decisions an initiator has taken.

Managing rather than giving room

A second perversion is that processes are made manageable with the help of project-type techniques. In such cases, a process is designed whose aims, preconditions, budget and planning are so tight, that only limited room is available for consultation and negotiation. It also follows the sequence of a project: a problem exploration phase is fixed, an aim is defined in the next phase, followed by a phase of information gathering, after which a decision is taken, which is implemented and evaluated in a following phase. The process follows the iron logic of project management, albeit parties 'are involved' in every phase, within predetermined preconditions. The process is thus cast in the mould of the project: those who have joined a process can move in only one direction (to the next phase) and have limited degrees of freedom.

> Much of the disappointment about interactive decision-making processes can be reduced to the project-type format of these processes. First, a number of concrete preconditions is defined, then the process approach is cast in the mould of a project approach. This leaves the actors involved hardly any room, although the use of the correct process language may suggest it.[31]

2.8.2 *Everything is a process*

The second risk is that the process fully replaces the competing management styles (substance, command and control, project management and single-value orientation). As we explained earlier, this thinking is too simple.

Process management leads to sluggish decision making

A major risk of process management is that it leads to slow and sluggish decision making. If many parties are involved in the decision making, there are many possibilities for these parties to block this decision making. If a manager indicates that he attaches importance to support and therefore organizes a process, this may even be an incentive for the parties to delay the process.

- If all parties are involved in the decision making and no decision is taken before there is support, it will pay for the parties to block the decision making. After all, so long as there is no support, there is no decision. A process approach thus gives rise to its own strategic behaviour.
- The mere fact that a manager organizes a process, is no reason yet for the parties involved to participate in it actively. They do not always have an interest in this process or in a prospective decision. A process approach thus leads to reactivity.

These two factors can delay the decision making immensely.

> Suppose a process approach is opted for too early in the decision making. On the one hand, the Maasvlakte proponents will join this process. They send representatives who have a mandate. Opponents have much less interest in this process. Conceivably, they will send representatives without a mandate or with a limited one, who have no interest in the progress of the decision making. For these opponents it pays to constantly block or hamper the progress of the process, causing the process to achieve the opposite of what is envisaged; no support and so no progress in decision making. Instead, it is delayed and blocked.

It is here that command and control become options again. Command and control can play a prominent role in preventing sluggish decision making.
A minister who negotiates with the business community about a partnership agreement and at the same time threatens to introduce unilateral regulations is likely to have a better partnership agreement than without the threat of hierarchy. Command and control is *not* used then to enforce a *substantive* change (the classic management style), *but* to enforce and boost a change *process*. In relations between states, negotiating while exerting pressure at the same time is sometimes referred to as 'bulldozer diplomacy'. Command and control and process each lead to resistance and sluggishness respectively. Proper process management means linking command and control and process intelligently (also see chapter 7).[32]

Process management impoverishes decision making

Process management – the thinking goes – enriches decision-making. A major risk, however, is that it rather impoverishes decision making.
First, if many parties have to be involved in the decision making, the results have to further the interests of these parties. The chances are that these results will be of below-average quality: they are a colourless and pale compromise.
Second, there is the risk of horse trading. Losers are more or less compensated at random (for example when, at one stage, they block the decision making), causing the eventual decision to consist of a package of issues that show no connection as to their substance.

> There is a difference between bargaining and negotiating. Bargaining is horse trading: the random coupling of subjects to reach a deal. Negotiating also implies coupling a number of subjects, but the parties then have as their criterion that this package must be consistent as to substance and must also lead to synergy and enrichment.

Third, enrichment is the product of a clash of ideas. The risk for a manager who is fascinated by the process approach is that he just organizes the process and confines his own role to managing this process. If enrichment is the product of a clash of ideas, such a process-managerial attitude is insufficient. Management without substance does not inspire, does not provoke opposition and thus cannot produce enrichment.

> Suppose the Rotterdam authorities in the Maasvlakte case opt for the visionary line: the Maasvlakte plan is presented to the public at large as *the* solution for future problems. The consequences are easy to imagine.
>
> - There will be considerable resistance, which can only be suppressed by expensive compensation.
> - There will be doubt about the use and the need.
> - Competing options will appear, which may even show rapid innovation (e.g. the bargees and the Betuwe freight railway line), causing doubts about the use and the need to grow.
> - Nonsensical arguments appear, such as: "Wishing to see such a strategic decision based on concrete and measurable arguments would be downright narrow-minded. What we need is vision";[33] Vision to justify the evasion of argumentation.
> - At best, a narrow victory for the proponents, causing decision making about projects of this kind and mutual relations to be mortgaged for a long time.
>
> A process approach is an alternative. Do not choose a solution too early, but define a number of functionalities or problems (for example, how to maintain the level of the functions of the area: the economy, the environment and nature, traffic and transport). Then start up a process to generate solutions that will find support. We do not have to be highly optimistic about the results. There is no substantive point of reference for parties to join the process. This will only evolve when a government formulates a vision of the problems in the relevant port. Such a vision creates movement and a clash of ideas. This will not be an integrated ready-made view, but a vision that makes clear to the parties that a direction is being chosen and that also offers them room to participate themselves in choosing that direction. The function of vision is to initiate a proper process rather than generate a binding substance.

Thus, process management is not opposed to substantive management, but interrelated with it. This leads to a highly subtle game: pose a view, propagate it vigorously and at the same time be prepared to adapt it and be willing to learn in a process. We will revert to this in chapter 8.

Notes

[2] See: J.A. de Bruijn, "From steering to process", in: R. J. in 't Veld (ed.), Steering illusion & disillusionment, Utrecht, 1999, pp. 52-68 (in Dutch).

[3] See: M. Douglas and A. Wildavsky, *Risk and Culture*, Los Angeles, 1982; M. Hisschemoller, *Democracy of problems*, Amsterdam, 1993 (in Dutch); H. van de Graaf and R. Hoppe, *Policy and politics*, Muiderberg, 1989 (in Dutch).

[4] J.A. de Bruijn, E.F. ten Heuvelhof and R.J. in 't Veld, *Process management: Decision making about the environmental and economic aspects of packages for consumer products*, Delft, 1998 (in Dutch); J.A. de Bruijn, R. van Duin and M.A.J. Huijbregts, in: J. Guinée et al., *LCA, An Operational Guide to the ISO-standards*, Dordrecht, 2002.

5 De Bruijn, Ten Heuvelhof and In 't Veld (1998).

6 Table copied from: De Bruijn and Ten Heuvelhof (2000); also see about networks and decision making: G.R. Teisman, *Complex decision making: a pluricentric perspective of decision making about spatial investments*, The Hague, 1992 (in Dutch); J. A. de Bruijn, *Processes of change*, Utrecht, 20001992 (in Dutch); D. Chisholm, *Coordination without Hierarchy*, Berkeley, 1989; P. Kenis and V. Schneider, "Policy Networks and Policy Analysis: Scrutinizing a New Analytical Toolbox", in: B. Marin and R. Mayntz (red.), *Policy Networks, Empirical Evidence and Theoretical Considerations*, Frankfurt am Main, 1991, pp. 26-59, 34 *ff*; also see: H. Willke, *Systemtheorie: Eine Einfuehrung in die Grundprobleme der Theorie Sozialer Systeme*, Stuttgart, 1993, pp. 236 *ff*

[7] See for example: R. Fisher and W. Ury, *Getting to Yes: Negotiating Agreement without Giving in*, Boston, 1981.

[8] De Bruijn, Ten Heuvelhof and In 't Veld, Process Management: Decision Making about the environmental and economical issues of packages for consumers, Delft, 1998.

[9] About strategic behaviour: Th. C. Schelling, The Strategy of Conflict, London, 1960; H. Riker, The Art of Political Manipulation, 1986; R. Axelrod, The Evolution of Cooperation, 1984; G. Egan, Working the Shadows Side: A Guide to 'Positive behind the Scenes' Management, San Francisco, 1994; D. Boddy and N. Gunson, Organisations in The Network Age, London, 1996; J. Boston (red.), The State under Contract, Wellington, 1995; A. Dixit and B. J Nalebuff, Thinking Strategically: The Competitive Edge in Business Politics and Every Day Lives, New York, 1991.

[10] J. A. de Bruijn, M. Kuit and E. F. ten Heuvelhof, *Sport 7*, Rotterdam, 1999.

[11] This form of argumentation is frequently found in the planning literature. See for example P. Healey, "Collobarative Planning in a Stakeholder Society", in: *Town Planning Review* (1998), pp. 1-21.

[12] Teisman (1992); W.N. Dunn, *Public Policy Analysis: An Introduction*, Englewood Cliffs, 1981; N. Crozier and E. Friedberg, *Actors and Systems*, London, 1977; M. D. Cohen, J.G. March and J.P. Olsen, "A Garbage Can Model of Organizational Choice", in: *Administrative Science Quarterly* (1972), pp. 1-25; A.G. Jordan and K. Schubert, A Preliminary Ordering of Policy or Network Labors, in: *European Journal of Political Research* (1992), pp. 7-27; A.G. Jordan, Sub-governments Policy Communities and Networks, in: *Journal of Theoretical Politics* (1990), pp. 319-338; W. van den Donk, The arena in tables, Tilburg, 1998 (in Dutch).

[13] Teisman (1992).

[14] R. E. Quinn, *Beyond Rational Management*, San Francisco, 1998.

[15] See for example: M.A. Maidique and R.A. Hayes, The Art of High Technology Management, in: *Sloan Management Review* (1984), No. 25, pp. 19-31; J. C. Collins and J.I. Porras, *Built to Last: Succesful Habits of Visionary Companies*, New York , 1997; K.S. Cameron, "Effectiveness as Paradox: Consensus and Conflict in Conceptions of Organizational Effectiveness", in: *Management Science* (1986), pp. 539-555.
[16] See about this: R.J. in 't Veld, *Northern Lights*, The Hague, 1997 (in Dutch).
[17] See: R.J. in 't Veld, *Playing with fire*, The Hague, 1995 (in Dutch)..
[18] R.J. in 't Veld, "The steering illusion of the purple coalition government", in: Frans Becker et al. (red.), Seven years of purple coalitions, Amsterdam, 2001, pp. 164-196 (in Dutch).
[19] De Bruijn, *Processes of change*, Utrecht, 2000, pp. 45-47.
[20] E.F. ten Heuvelhof, Standards of behaviour for governments in horizontal structures, The Hague, 1993 (in Dutch).
[21] Cohen Commission, Market and government, The Hague, 1997 (in Dutch).
[22] R.J. in 't Veld, Half a loaf is better...? An essay about the government on the market, The Hague, 2001 (in Dutch).
[23] Directorate General for Public Works and Water Management North-Holland, Handbook for the open-plan process, s.l., 1996 (in Dutch).
[24] Bosch, F.A.J. van den, S. Postma, Strategic Stakeholder Management: A Description of the Decision-making Proces of A Mega-investment Project at Europe's Biggest Oil Refinery: Shell Nederland Raffinaderij BV Erasmus University/Rotterdam School of Management, *Management Reports Series* no. 242, Rotterdam 1995
[25] D. Vaughan, The Challenger Launch Decision: Risky Technology, Culture, and Deviance at NASA, Chicago, 1996.
[26] G.R. Teisman, Steering through creative competition, Nijmegen, 1997; Mobilising space through cooperative management, Rotterdam, 2001 (in Dutch).
[27] De Bruijn, Ten Heuvelhof and In 't Veld (1998).
[28] Giddens speaks of 'dialogic democracy', which supposedly leads to an increase in social reflexivity. A. Giddens, *Beyond Left and Right: The Future of Radical Politics*, Stanford, 1994.
[29] Giddens (1994) mentions as one of the arguments that this creates confidence between the parties.
[30] Part of the following argument has been copied from: De Bruijn, "From steering to process", in: *Steering illusion & disillusionment* (1999), pp. 52-68.
[31] Marc Chavannes, The sluggish state, Amsterdam, 1994 (in Dutch).
[32] J.A. de Bruijn and E. F. ten Heuvelhof, *Networks and Decision Making*, Utrecht, 2000
[33] Quoted through the *Financieel Dagblad*, 10 November 1998

PART II PROCESS ARCHITECTURE

3. DESIGNING A PROCESS

3.1 INTRODUCTION

This chapter focuses on making a process design: rules of the game that the parties involved will follow to reach a decision. First, section 3.2 deals with the four key requirements a process design has to satisfy. We will call them the core elements of a process design in the rest of this book. A good process is an *open* process, in which parties' core values are protected, which has sufficient *incentives for speed* and offers sufficient guarantees for the *substantive quality* of the results. We will then translate each of these core elements into a number of design principles (sections 3.3 to 3.6), which together constitute a series of criteria a process design should meet.

3.2 THE FOUR CORE ELEMENTS OF A PROCESS DESIGN

The four core elements of a process design we distinguish are based on the following reasoning.

- *Openness* Process management means that an initiator does not take unilateral decisions, but adopts an open attitude. Other parties are allowed to participate in steering the decision making and can therefore also indicate what items interesting to them should be placed on the agenda.
- *Protection of core values* Openness is not always attractive to parties invited to participate in a process. There is a risk that they will be unable to advance their own interests sufficiently. As a result, they might be dissatisfied with the outcome, while it is difficult for them to withdraw from the process at that stage. This is why the second category of design principles springs from the idea that the parties committing themselves to a process must be given sufficient protection of their core values. They must be certain that their core values will not be harmed, regardless of the outcome of the process. This makes the process a safe environment for them.
- *Speed* The first two core elements offer insufficient guarantees for a good decision-making process. If participants opt for open decision making (core element 1), in which parties' key interests are protected (core element 2), there is a substantial risk that no decisions will be taken, even though consultations and negotiations take place. The only outcome may be sluggish processes, which will never produce a clear result. The third category of design principles concerns the need for the process to have sufficient speed and progress.

- *Substance* Parties participating in an open process (core element 1) should be given sufficient protection of their position (core element 2), while there should also be sufficient guarantees that progress will be made in the decision-making process (core element 3). In the fourth place, this progress should also meet the requirement of substantive quality. Forced by the sharp conflicts of interests, the parties may, after all, take decisions that are meagre from a substantive point of view, or even incorrect. The process should have enough substantive elements.

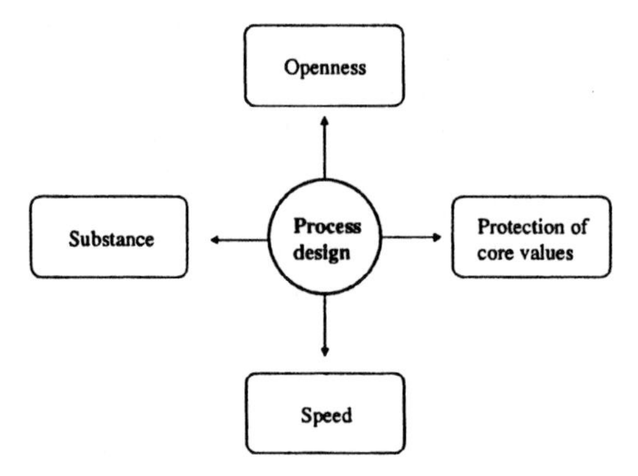

Figure 3. 1 Core elements of a process design

A process design will always have to take the four core elements into account and will always be a trade-off between them. A process without openness will be perceived as a disguised form of project management and command and control. A process not protecting parties' core values is highly unattractive to these parties. There is a risk that they will constantly delay the process out of distrust or even not join the process at all. The process will become sluggish if there are no tools to give it progress and speed. A process may produce meagre results if there are no procedures to create substance and quality is lacking. A process architect always has to make a trade-off between these core elements.

Process management is often associated with seeking support and contrasted with seeking rapid and efficient decision making. The value of openness then surpasses that of speed. This is a one-sided picture, because a different trade-off may also be made between the four core elements. For example, both in process design and process management, the value of progress may dominate in the trade-off.
Since the liberalization of telecommunication, former monopolists (the national post offices) have been negotiating with newcomers about interconnection rates. These are rates the newcomer must pay for the use of the network and/or the services of the

former monopolist (the ' incumbent'). If a dispute arises between an incumbent and a newcomer about the interconnection agreement during the negotiations, it may be submitted to a mediator, who may settle the dispute in consultation with the parties. The mediator can use a process design, which – in most cases – will contain strong incentives for cooperative behaviour. Parties tend to have a particular interest in a rapid process, enabling them to know what they can expect with regard to their own business operations. The substance of the outcome is less important then, because a good outcome that takes too long to achieve leads to a long period of uncertainty and harms business operations. [34]

Sections 3.3 to 3.6 describe the design principles. Table 3.1 summarizes the design principles for the sake of clarity.

Openness	1. All relevant parties should be involved in the decision-making process. 2. Substantive choices should be transformed into process agreements. 3. Both the process and its management should be transparent.
Protection of core values	4. Parties' key interests should be protected. 5. Parties should commit themselves to the process rather than to the result. 6. Parties may postpone their commitments to decisions made. 7. The process should offer participants and exit option.
Speed	8. The process should create prospects of gain as well as incentives for cooperative behaviour. 9. The participants in the process should have commitment power. 10. The process should have an environment, which is used to speed it up. 11. Conflicts should be transferred to the periphery of the process. 12. Command and control can be used as an incentive to speed up the process.
Substance	13. The process should prevent the process-drives-out-content mechanism; the roles of experts and stakeholders are both bundled and unbundled. 14. The process should move from substantive variety to selection.

Table 3.1 The design principles

3.3 DESIGN PRINCIPLES LEADING TO OPEN DECISION MAKING

3.3.1 All relevant parties should be involved in the decision making[35]

The first design principle is that all relevant parties should be involved in the decision making. A key question is, of course, who the relevant parties are. We will list four considerations.

- *Parties that have blocking power in the decision making.* These are parties that can block the decision making. Involving these parties in a process may keep them from exercising their blocking power, which would frustrate the decision making. Of course, the consideration here is whether this mechanism will actually occur. Parties will exercise their blocking power during the process if it is insufficiently attractive to them. This consideration may then be a reason not to invite the party in question. Such a decision will of course depend on the amount of blocking power a party has. If it is absolute, excluding that party makes little sense.
- *Parties that have productive power in the decision making.* These are parties that have the means to actually implement a decision (or part of it). Examples include money, powers, relations, physical resources or expertise. In many cases, these means may enrich the decision making. A party that has a great deal of expertise may contribute it to a process about expanding an airport, for example. Such a party has an insight into the latest technical possibilities of reducing noise, laying down approach routes intelligently or promoting an efficient use of airspace; such knowledge can break a deadlock about these items between negotiating parties.
- *Parties that have an interest in the decision making* without being able to contribute to the solution of a problem or block a solution. In the above example, there may be particular municipalities situated far from the airport, as a result of which they have hardly any resources (in procedural law) to influence the decision making. If these municipalities nevertheless suffer noise nuisance, the airport manager may involve them in the decision making on moral grounds.
- The last example shows that there are considerations other than those of power. There may be *important moral considerations to involve particular parties in the decision making.* In almost every process there are parties whose position is negligible as regards power, but that might suffer as a result of a potential decision.

It should be pointed out, however, that all these are just considerations: the mere fact that a party meets this criterion does not mean that it must be invited. The differences between parties may, of course, lead to differences in the way they are involved in a process. A process may have a number of phases, for example, and – deliberately – fewer parties are invited in an opening phase than in later phases. Different roles may be defined as well: particular parties are involved in the decision making, other give obligatory advice, others again participate in the process as experts, et cetera.

3.3.2 Substantive choices should be transformed into process agreements

The second design principle is that as few substantive choices as possible are made before the start of the process. A transformation from substantive choices to process agreements takes place. Although the substantive items are inventoried first, it is then merely indicated how the decision-making process will take place

3.3.3 The process and its management should be transparent

Consequently, it is very important that the design for a decision-making process should be transparent. Transparency means that the course of the process is clear to parties, that it is clear to them how their interests will be protected, what decision rules will apply, and – of course – who will be involved in the process. Occasionally, the metaphor of the itinerary is used here: it should be clear what the last leg of the journey will be, where passengers can board and get off, what will happen if there are any delays, et cetera. Transparency means that the parties can check whether the process is sound and whether it offers them enough opportunities.

An opaque process, in which parties do not know what the process agreements are, is a breeding ground for mutual distrust and –as a consequence – for mutual conflict.

The behaviour of the process manager also has to meet the transparency requirement. The process manager has the role of an independent facilitator. Of course, the process manager should concentrate on the process rather than the substance: strong substantive views may be regarded as substantive prejudices by one or more parties, and thus as an infringement of the process manager' s independence.

3.4 DESIGN PRINCIPLES PROTECTING PARTIES' CORE VALUES

3.4.1 Parties' core values should be protected[36]

Openness means that the parties joining a process must be allowed to influence future decision making; the impression that they will get hooked in a trap should be avoided. This particularly applies to parties that are difficult to persuade to join a process. These parties should not be led to believe that their participation implies that they will get caught in a decision-making process, that the envisaged decision has in fact been taken already and that they partly justify it by their participation. One way to prevent this is to offer the parties protection of their core values: they may then be sure that, as regards these values, they will not be forced to adopt a certain behaviour against their will. Core values are an organization's 'raison d'être', failing which they cannot exist. Core values concern the essence of an organization and therefore exceed the importance of one single issue about which decisions were taken. On the contrary, core values concern almost all issues an organization deals with. A core value should not be confused with a stance an organization adopts in a decision-making process. The core value surpasses the process and the single point of view.

3.4.2 Parties should commit themselves to the process rather than to the result

An important aspect of the protection of core values is that parties cannot be asked to commit themselves ex ante to the result of a process (even though it is both fair and sound in its design); all that may be asked for is commitment to the process. This creates room for the parties, making the process a safe environment, in which the risk of core values being harmed is limited

3.4.3 Parties may postpone their commitments to decisions made

A decision-making process tends to consist of a large number of subdecisions, which will result in a final decision at some stage.
One important aspect of room for and protection of core values is that a party does not have to commit itself to these subdecisions during the process. This is because the complexity and the importance of the problems for which a process design is made may cause parties to fear such commitment. If parties have to commit themselves to particular decisions early in the process, they may perceive this as passing a point of no return. This makes it difficult to go back on earlier positions later in the process without harm. Commitment to subdecisions will reinforce the perspective of the process-as-a-trap, causing parties sooner to feel that their core values are being harmed, which may frustrate the course of the process.

3.4.4 The process should offer the participants an exit option

An important design principle is that a process has exit rules: rules allowing parties to exit the process while it is in progress. For example, it may be included in the process agreements that the parties may decide after some time whether they wish to continue their participation in the process. This lowers the barrier for particular parties to join the process. Such an exit rule also greatly reduces the risk of participation for an individual party. A party is allowed to leave the process before any final decisions are taken. This removes the trap perception.
It should be pointed out, however, that participation in the process will, after some time, be so attractive to the parties that exiting is no longer an option (also see chapter 6).

3.5 DESIGN PRINCIPLES GUARANTEEING THE SPEED OF THE PROCESS

3.5.1 The process should create prospects of gain and incentives for cooperative behaviour[38]

The parties committing themselves to the process design are mutually dependent, and there tend to be differences between them that are difficult to bridge. The natural behaviour of these parties tends to lead to mutual discord and mutual conflict. A design for a decision-making process will therefore have to contain incentives for cooperative and disciplined behaviour. The chief incentive for cooperative behaviour by parties is the prospect of 'gain'. This means that parties must be convinced that the process will be, and will continue to be, attractive enough for them to participate wholeheartedly.

Commitments to subdecisions may be postponed during a process (see above). There should always be sufficient room for the parties involved. A cynical analysis of this might be that decision making is constantly postponed and that the process produces no results. From a perspective of process management and creating win-win situations, however, something entirely different is the matter. There continues to be a prospect of gain for the parties, and an incentive to continue participating in the process is created because room is offered for future decision making.
It is important here that the 'gain' for the various parties should not be paid too early. There is no longer any incentive for a party to cooperate once it has received the gains it has made. There will be a risk of opportunistic behaviour then: the party in question either withdraws from the process or no longer behaves cooperatively. Accordingly, the process architect should ensure maximum possibilities for making gains at the end of the process.

3.5.2 The participants in the process should have commitment power

The second design principle concerns the question what people should participate in a process. In accordance with this design principle, a process should have participants who have authority and commitment power: the power to commit their rank and file to the decisions made in the process. Three arguments for this design principle all indicate that commitment power facilitates the decision making.

- The first argument is that it increases the process's external authority as well as its aura. As we set out earlier, this is an important precondition for the progress of the process as well as for the decision making.
- Secondly, if the representation is too lightweight, commitment power can often be obtained only by a strong formalization of the relation between the representative and the organization represented: a great deal of consultation is required and mandates are detailed. This may seriously hamper the progress of the process. Commitment is far more a matter of course if there is a heavyweight representation.
- The third argument is that, if necessary, an authoritative representative can distance himself to a certain extent from those represented. This is also important in processes of this kind, in which parties may have to make concessions, accepting which may be difficult for them.

3.5.3 The process should have an environment, which is used to speed it up

Every process has its environment: parties that do not participate in the process, but that do have an interest in the process or its outcome. In terms of the game metaphor: the players are surrounded by spectators on the stands. Although they do not take part in the game, they can exercise some influence. Utilizing this environment may benefit the progress and speed of the decision making. This is called environment management.

> The Board of Management of two merging divisions may deliberately position itself at some distance from these merger partners. The divisions should take decisions about the merger in mutual consultation. The Board of Management forms part of the divisions' environment rather than being directly involved in the merger. The management of the divisions may lay down the decision-making process for the merger in a protocol or a number of process agreements. The Board of Management may approve these process agreements without interfering in their substance, and may threaten to intervene if the process stagnates or one of the divisions tries to evade the process. The Board of Management's approval of the process may counteract potential criticism of this process by the participating divisions (since this might cause delay). Pressure from the Board of Management may also promote the

decision making. The management of the divisions may make active use of it by provoking Board of Management interventions at certain stages.

3.5.4 Conflicts should be transferred to the periphery of the process

A process is designed in view of potentially discordant relations between the participating parties. A process approach poses an evident, major risk: the parties are brought together, as a result of which conflicts are fought out with great fierceness. Conflicts may consequently stagnate the process. Here, the image of fighting cocks is used occasionally: a process may be designed to bring fighting cocks closer together, but once the process has started there is a risk that they will wage a life-and-death struggle, fiercer than ever before.
The process will have to have a number of provisions to prevent too many conflicts arising between the parties during the process. Also, many processes have a layered organizational structure. There is a core, around which there are a number of skins. In many cases, there is a steering committee-project group-working group structure, for example. The key decisions are taken in the steering committee. They are prepared in one or more project groups, which may delegate particular day-to-day activities to a working group. Such a structure allows conflicts to be located in the periphery of the process.

> Conflicts may arise in project groups and working groups without posing too much of a risk, because they do not, or do not directly, burden the representatives in the steering committee. The members of the steering committee may use the positive effects of conflicts–more information, a better understanding of the differences–in their decision making.
> A special role tends to be reserved for external experts, who may be consulted at particular moments. This creates an interests circuit and a circuit for experts. The layering of the organization thus increases. Here, too, these experts can play a role in solving or mitigating conflicts.

This mitigation of conflicts is necessary because it prevents the parties' representatives being faced with too many conflicts. This is because they tend to have a limited conflict-handling potential.

3.5.5 Command and control can be used as an incentive to speed up the process

Chapter 2 contrasted process management with the management style of command and control. However, to suppose that command and control can play no role whatsoever in process design and process management is too simple. Particular forms of command and control may be an incentive for parties to join the process and adopt a cooperative attitude in it; it is thus a driver for process management. It is equally important that parties may become more amenable to command and control during the process, for example because they stand to gain or because they learn that consultation alone produces no results. This fact may be used in designing a process.

Suppose, for example, that a formally competent body does not want to take a decision about a subject because it expects strong opposition from the parties involved. This body may then make an agreement to the effect that these parties will form a judgement about the subject in question in a process of consultation and negotiation. A following agreement may be that, if there is consensus between parties, this body will follow this judgement as much as possible in exercising its powers. The agreement also says that the competent body will use its own discretion if there is any discord. In other words, it can use its powers unilaterally in the event of discord and therefore steer by command and control. The preceding process is necessary, however, to make parties accept this command and control.

3.6 DESIGN PRINCIPLES GUARANTEEING THE SUBSTANCE OF THE PROCESS

3.6.1 The process should prevent the process-drives-out-content mechanism; the roles of experts and stakeholders are both bundled and unbundled[39]

The above says little about the substance of the process. Although the process architect is not a substance expert, he cannot, of course, ignore the substance. A process without substance is empty. Substance can never determine a decision-making process, however. After all, one of the justifications for a process design is that it is impossible to solve problems on the basis of objective information.

A decision-making process is vulnerable when drifting too far away from the substance. The impression may be created that it is dealing with nothing at all. Process drives out substance. Although one party will be more tolerant as to the distance between process and substance than another, there is a kind of *line we dare not cross*.

For the process architect, all this means that he should ensure that the process is sufficiently substantive. The process should be structured so as to allow relevant substantive insights to play a role within the process.

The process architect deals with this requirement in a process-oriented way. He may do so, for example, by including substance experts in a decision-making process in addition to stakeholders. They may facilitate the decision-making process with their substantive knowledge. They can unmask fallacies in the stakeholders' arguments, trace inconsistencies, separate sense from nonsense, perform sensitivity analyses, et cetera.

Making a clear distinction between experts and stakeholders makes clear to everyone who plays what role. It also prevents experts 'changing colour'. If these experts become too closely involved in the conflicts of interests between the parties in the process, partly as a result of an unclear division of roles, there is a risk that they will no longer be able to play their independent role, but will become a party with interests in the process.
Such a distinction also poses the risk of unbundling substance and process too strictly (also see chapter 2 about the risks of unbundling substantive rationality and process rationality).
This design principle counteracts this risk by indicating that unbundling roles must be accompanied by a clear bundling of activities. Bundling means that structured relations are maintained from a position of independence. This allows decision makers to question the substance experts about the scholarly character of their research findings or views: what assumptions they are based on, what data are used, what the system boundaries are, et cetera. Researchers can test the scientific tenability of stakeholders' views. Which views can stand the test of scientific criticism and which cannot?

3.6.2 The process should move from substantive variety to selection

This does not mean that scientists or experts will come up with final answers. They facilitate the decision making rather than determining it. The outcome of the test of scientific criticism may be that some views are correct and others are not, with a large number of views falling within a bandwidth of what is tenable from a scientific point of view.
As regards the views that fall within this bandwidth, the process should move from variety to selection (see chapter 2). First, a variety of views should be submitted, after which selection can take place. Selection made too early or based on a limited variety will not usually be authoritative.
The design principles presented here may help the process architect to design a process. Chapter 3 describes which activities should be performed for this purpose. We will revert to each of these design principles in part III. We will then discuss them not from the perspective of the process architect, as we did in this part, but from that of the process manager.

Notes

[34] J.A. de Bruijn, E.F. ten Heuvelhof and H.I.M. de Vlaam, *Interconnection disputes*, in: *ITeR series* (1997), No. 8, pp. 165-265, Alphen a/d Rijn.

[35] Innes (1996), pp. 460-472; F. Fischer, *Citizens, Experts and the Environment*, London, 2000; J. Bohman, *Public Deliberation: Pluralism, Complexity and Democracy*, Cambridge, Mass., 1996; S.R. Arnstein, "Eight Rungs on the Ladder of Citizen Participation", in: S.C. Edgar and B.A. Passet (ed.), *Citizen Participation: Effecting Community Change*, New York, 1971, pp. 69-91; E.G. Guba and Y.S. Lincoln, *Fourth Generation Evaluation*, Newbury Park, 1989; I.S. Mayer, *Debating Technologies. A Methodologica Contribution to the Design and Evaluation of Participatory Policy Analysis*, Tilburg, 1997; J. Edelenbos, *Process in form: Overseeing the process of interactive policy making about local spatial projects*, Delft, 2000 (in Dutch).

[36] C.W. Moore, The Mediation in Process: Practical Strategies for Resolving Conflict, 1996; Ph.W. Kheel and W.L. Lurie, The Keys to Conflict Resolution, Proven Methods of Settling Disputes Voluntarily, 1999.

[37] R.J. In ' t Veld, J.A. de Bruijn, E.F. ten Heuvelhof et al. *A process standard for the implementation of the Packaging Covenant*, Rotterdam, 1992 (in Dutch).

[38] N. van Baren, Plan-hierarchical solutions: A source of social resistance, Amsterdam, 2001 (in Dutch); C. Lambers, D.A. Lubach and M. Scheltema, *Speeding up legal procedures for large projects: Preparatory study for the Netherlands Scientific Council for Government Policy*, The Hague, 1994 (in Dutch); A.J.F. Bruning, *Large projects in the Netherlands: An analysis of the length of 20 decision-making processes*: Preparatory study for the Netherlands Scientific Council for Government Policy, The Hague, 1994 (in Dutch); W.M. de Jong, *Institutional Transplantation: How to Adopt Good Transport Infrastructure Decision-making Ideas from Other Countries*, Delft, 1999; Dixit and Nalebuff (1991); A. Sparks, *Tomorrow is Another Country: The Inside Story of South Africa' s Negotiated Revolution*, London, 1995, p.180; de Bruijn (2000); P.C. Stern and H.V. Fineberg (ed.), *Understanding Risk Informing Decisions in the Democratic Society*, Washington, 1996.

[39] S. Jasanoff, *The Fifth Branch: Science Advices as Policy Managers*, Boston, 1990; *Science and Public Policy* (1999), vol. 26, No. 3, theme issue about scientific expertise and political responsibility; W.M. de Jong, *Institutional Transplantation: How to Adapt Good Transport Infrastructure Decision-making Ideas from Other Countries*, Delft, 1999; Edelenbos (2000); M.L. Miranda et al., "Informing Policymakers and the Public in Landfill Siting Processes", in: *Technical Expertise and Public Decisions*, Institute of Electrical and Electronic Engineers, Princeton, 1996.

4. THE PROCESS ARCHITECT IN ACTION: MAKING A PROCESS DESIGN

4.1 INTRODUCTION

The design principles may help in making process agreements, but the question is of course how such agreements come into being. In this chapter, we will answer these question as follows.[40] In section 4.2, we will set out that process agreements always result from some form of negotiation. Standardization of process agreements is almost impossible, because the configuration of actors and the substance of the problems to be solved will always differ. In section 4.3, we will discuss an important condition for process management: the parties involved must have a sense of urgency telling them that they need each other to solve particular problems. Process management is unlikely to succeed without it. Section 4.4 then lists the main activities that a process architect performs and that result in a process design. We will conclude with a discussion in section 4.5 positioning the process approach in relation to procedural rationality, consensus building and interactive decision making.

4.2 THE PROCESS DESIGN AS A RESULT OF NEGOTIATION

An important condition for the success of a process design is that it should be attractive to each of the parties involved: they should be convinced that the design offers them a fair chance of influencing the decision making and that it will not harm their core values.
Now it will be difficult for one of the parties (or for an independent third party) to draw up an attractive process design unilaterally, especially if there are serious conflicts of interests between the parties. The conclusion is an obvious one: an attractive process design, of which all parties involved are the joint owners, can only come into being if these parties can participate in shaping it. In other words, the process design is also the outcome of a process.

In the Netherlands, all students receive a free-of-charge annual season ticket for public transport. It allows them to use public transport on particular days of the week either free of charge or at reduced fares.
Students receive this ticket as part of the financial contribution granted them by the Ministry of Education to fund their studies. For this facility, the Ministry of Education pays a sum to the public transport companies, based on an agreement between the Ministry of Education and the public transport companies. This agreement has to be renewed from time to time.

The Ministry of Education and the public transport firms have already negotiated twice about a new season ticket for students. In these negotiations, the public transport firms make an offer to the Ministry of Education, stating their price for a week-weekend ticket for a particular category of students. The negotiations then start. In them, the Ministry of Education is interested in a transparent quotation: it should be clear what cost items form the total amount, what the costs of each component are and how many kilometres students actually travel on the ticket. The public transport firms have an interest partly conflicting with that of the Ministry: they want to protect the confidentiality of their business information. It will be clear that it is difficult to make a standard process agreement about this. In negotiations, parties will have to reach a procedural agreement that satisfies these conflicting interests. Such an agreement may be, for example:

- Parties may ask each other for factual information, which they will make available where possible.
- If the public transport firms regard particular information as confidential, they will state the reasons. The Ministry of Education will form a judgement about it. If the parties disagree about confidentiality, they are entitled to state this in the (public) document containing the final result.
- If the public transport firms regard particular information as confidential, the Ministry of Education reserves the right to gather the information in question by other means, for example by research.

We are not concerned here with this concrete illustration, but with the fact that such agreements can hardly be devised beforehand and can only come about in negotiations. Even if they can be devised beforehand, it would still be wise to make them the outcome of a process. Only then will the parties feel themselves the joint owners of the agreements.

Incidentally, if no process agreements are made, there is every risk that discord will arise during the negotiations about the information the public transport firms provide and about whether the public transport firms have a duty to provide further information. It is highly questionable whether, in the heat of the talks, any reasonable negotiations are possible at all about the rules of the game on this point.

Negotiating about a process design would seem inefficient at first sight, particularly for an initiator who initially believed that he could implement a change unilaterally. After all, this initiator will first have to accept his dependence on others and enter into a process of consultation and negotiation with them. To conduct these substantive negotiations, a process design is necessary. It will then be found that negotiations also have to take place about the rules of the game in the process design: procedural negotiations precede substantive negotiations. Although this may surpass the pain limit of the party in question, there are four positive effects of procedural negotiations:

- Such negotiations imply that parties can learn about each other: about their interests, sensitivities, core values, room for solution, et cetera. If this learning process took place during the substantive negotiations, it might seriously disturb them. The advantage of negotiations about process agreements is therefore that this learning process takes place during the procedural negotiations without causing such disturbance.
- In fact, negotiating about a process design means that the substantive negotiations have started. This is because the proposals for process agreements that parties make tend to be prompted by the substantive issues they want to put up for discussion. These substantive issues will be discussed, however, without parties actually having to make substantive agreements. This enables them to learn about the substantive agenda of the negotiations.
- Parties also learn about the need to conclude process agreements. If, during the procedural negotiations, they find that parties' interests differ sharply, that agreement is not a matter of course and that the relation between parties generates conflicts, the risk of internalization of the process agreements as well as the obligation to respect them as a party will increase. This, too, is important, because in the phase of substantive negotiations (so when there is a series of process agreements) a party may be tempted to ignore process agreements when it suits that party from a substantive point of view.
- Process agreements are more likely to succeed, of course, if they are the joint product of the parties involved. If not, it is relatively easy for a party to distance itself from the process agreements during the substantive negotiations if it feels that these process agreements favour one of the parties.

4.3 THE NEED FOR A SENSE OF URGENCY

Process management can only succeed if the chief stakeholders have a sense of urgency.[41] This means that these parties are convinced that there are problems that need to be solved and that they can only be solved by some form of cooperation.
Such a sense of urgency is a sine qua non for the success of a process. Without it, parties will hardly be prepared to negotiate about process agreements. If agreements are made nevertheless, they are unlikely to be respected, because they can easily be perceived as impediments if there is no sense of urgency.

All this means that the process approach may be introduced too early in a decision-making process. Decision making in a network has its natural dynamic: there is an initiative, a process of pushing and pulling between actors, which either leads to result at some stage or, instead, stagnates. In the latter case, a sense of urgency may develop among parties, which is a breeding ground for process agreements between parties. Process management is not an instrument to be used at one's own discretion. Momentum for process management should evolve in the decision making first.

> Particularly now that process management is gaining popularity and fame, process designs are already made at the start of the decision making in more and more cases. For example, an adviser finds that many parties are involved in an issue and that a process approach is therefore necessary.
>
> This is usually the wrong kind of efficiency: first, a sense of urgency needs to evolve among the parties, which may take some time. In chapter 2, we discussed the example of the Second Maasvlakte. Suppose Rotterdam defines lack of space as the problem and the Second Maasvlakte as the solution. Suppose the municipality then makes an actor scan and invites parties to participate in a process on the basis of it. The consequences are obvious.
>
> - The decision making is still far away, which makes it hardly attractive for parties to participate in such a process, causing them to adopt a reserved attitude.
> - Furthermore, the parties have insufficient knowledge and information – particularly about complex issues like the Maasvlakte – and they may not have formed an opinion yet. This also leads to reserve among the parties.
> - Parties may distrust the invitation. They do not share Rotterdam's perception of the problem and may regard the process as a means to commit them to this problem definition.
>
> A project manager may take a different view: the sooner decisions are taken, the better. A process manager may take the view that they should be taken at a late stage rather than at an early stage. At an early stage means that the process will get bogged down; at a late stage means that the sense of urgency has risen high and that parties are therefore prepared to invest a great deal in a process. In some cases, the best advice may be not to interfere with a decision-making process for a while, because the parties first have to become convinced that they cannot make any headway without cooperation.

Those who nevertheless start a process may almost be sure that it will die of sluggishness: parties fail to reach a process design or treat it carelessly if there is one at all. This will cause a risk. If, after some time, the experience of parties is such as to create the sense of urgency, they will have to cooperate in a process. In that case, however, there will be a *burden of the past*: parties have bad experiences with an earlier process, hampering cooperation in the process that is necessary now. Those who organize processes that are either ill-considered or too early will harm future processes from the very start.

4.4 THE PROCESS ARCHITECT IN ACTION: DESIGNING A PROCESS

In this section, we will describe the activities performed by a process architect and resulting in a process design. The following remarks are appropriate here.
First, we will present a complete list of activities to be performed. It is not always necessary, however, to perform all these activities to achieve a good process design. Either more or fewer of these activities are required, depending on the substantive complexity of the issue and the seriousness of the conflicts of interests. In some cases it would be opportune not to perform certain activities or only to perform them to a limited extent only.
Second, the order of activities is less compelling than the following list may suggest. A different order may be opportune in some cases; in other cases, iterations may be appropriate, an activity being repeated either once or several times.
Third, a process architect with a sense of public management relations knows that it may be wise not to make all activities explicit. We will illustrate this below in answer to the question how a process manager should structure an agenda.
Finally, a qualification about the list of activities is fitting. Performing these activities, combined with the correct use of the design principles, does not necessarily produce a good process design. Designing a process is not a mechanistic activity, the accurate performance of whose steps automatically leads to a good design. It takes more to design a good process. Rather than a skill, process design is almost an art. There is a touch of virtuosity about some designs. The process architect also needs an eye for the elegance of process arrangements, he should have a sense of managerial relations, know about administrative mores, about the *burden of the past* and about persons and characters. In addition, great linguistic skills and well-developed conceptual powers are indispensable to achieve a process design commanding managerial support.

Figure 4.1 describes the activities a process architect passes through to make a process design.

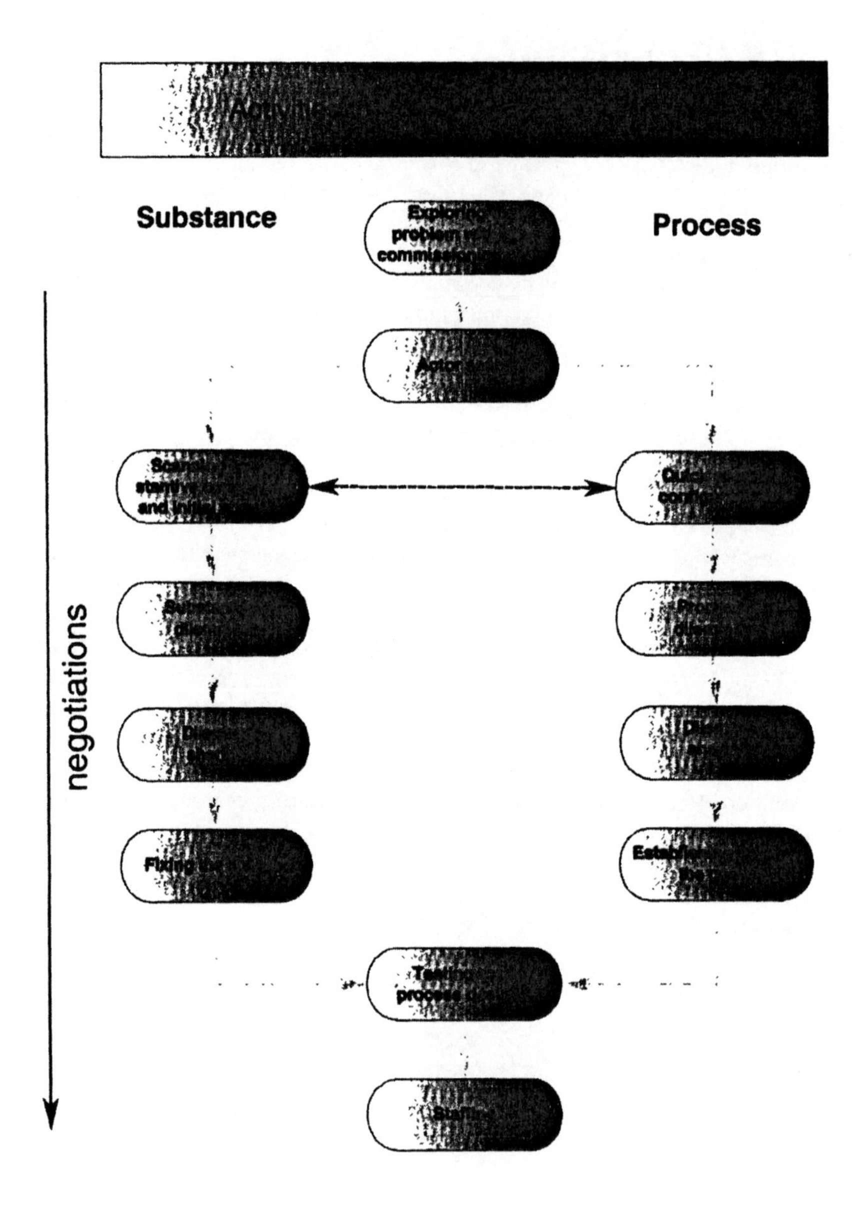

Figure 4.1 Activities for making a process design

Exploring the problem with the commissioning party

Although the essence of a process may be that it involves several actors, the start of a process tends to be a matter of one actor or a limited number of actors. This initiator or these initiators contact the process architect saying they want to realize something and that they are concerned about the course of the decision-making process so far. They feel the quality of the decisions is low and progress is slow, are concerned about support for the decisions or about managerial commitment. This initiator commissions the process architect to get the decision-making process afloat. In this opening phase, he performs the following activities.

- First of all, he makes an initial problem exploration in consultation with the commissioning party. Inevitably, this exploration is strongly coloured by the views of the commissioning party. The process architect should emphasize that this problem exploration is no more than a first exploration and that it is a precondition for a successful process that all relevant actors should endorse the final agenda.
- The next step is that the commissioning party and the process architect draw up an initial list of relevant actors.
- The third step in this first exploration with the commissioning party is to map the sense of urgency of the process. The commissioning party will almost certainly feel a certain urge to advance the decision making. Otherwise he would probably not have contacted the process architect. But does that also go for the other actors? Do they also feel the need to do something about the decision making, or do they feel no such urge, or do they perhaps rather value the stagnation as positive? The process architect should form an image here of the views of the other actors.
- If the process architect and the commissioning party conclude that the sense of urgency is uncertain, the following possibilities are open to them. Their first option is that they should jointly decide that designing and starting a process stands no chance of success under these circumstances. However, the fact that a process is either a non-starter or a failure may deteriorate relations between actors, thus placing a heavy burden on any following processes. Consequently, the second option is that the commissioning party and the process architect decide to postpone answering the question about the sense of urgency and find out in a following round what sense of urgency the actors themselves feel. For example, a sense of urgency may emerge when discussing the agenda, which contains more items than those mentioned by the commissioning party. The final option is that the commissioning party and the process architect explore the possibilities of increasing the pressure on the process so as to create a sense of urgency among the other actors as well.

4.4.1 Actor scan

The first exploration with the commissioning party being completed, the first scan of the relevant actors will follow. Actors will be selected in two ways. The first selection is made in consultation between the commissioning party and the process architect. This list is expanded during the scan. New actors, not mentioned initially by the commissioning party, but nevertheless important for the progress and quality of the process, will come up for discussion in the analysis of written documents and during interviews with actors. Having consulted the commissioning party, the process architect should also involve these actors in his scan.
The aim of the scan is to gather a number of details about each individual actor. The sources of the scan for an individual actor are threefold: an analysis of the written documents, an interview with the actor itself, and interviews with other actors about the reputation of the individual actor in question. The process architect gathers information about each actor as regards points of view, interests and core values, opportunities and threats, incentives and disincentives as well as multiformity. We will briefly explain this.

Points of view, interests and core values
The process architect maps each actor's points of view about an issue. Is the actor in question a proponent or an opponent of the commissioning party? Or is the actor keeping his options open? Does the actor mention conditions that must be met before he cooperates? How impassioned is his attitude towards the proposal?

The process architect discovers interests by gathering information about the background of the actor's point of view. Why is he a proponent or an opponent? What does he expect to reach by being one? In practice, it is often other actors that can explain why an actor takes a particular view rather than the actors themselves identifying the interest.
In the quest for interests, a distinction should be made between formal and informal interests. The actor itself will mention the formal interest behind its point of view. This interest refers to the organization's 'raison d'être'. Informal interests tend to be more difficult to formulate and may even be slightly banal, which does not make them less relevant to the process, however.
The process architect who has formed an image of an actor's points of view and interests is able to form a first impression of their 'elasticity'. What movements in the point of view are imaginable, given the actor's interests?
Special interests and interests of extraordinary importance to the actor fall under core values. Core values are more general rather than relating exclusively to an issue being fought over in the decision-making arena. They are vital to the actor,

however, and also apply to the issue under discussion. Core values have to be disclosed, because they need protection in special arrangements. Offering this protection makes it easier for actors to join a process (also see chapter 6).

Opportunities and threats
The process architect maps the opportunities each of the actors sees with regard to the issue the initiator wants to address in the process. It is also important here that a process manager should check what issues may be coupled to it so as to make a process attractive to the other parties. If parties are offered many opportunities in a process, this may be a reason for them to commit themselves to a process. The process architect also pays attention to the threats from an actor's perspective: what agenda makes an actor hesitant about joining a process?

Incentives and disincentives
The process architect should also map the incentives and disincentives for each actor. Incentives are the factors inciting an actor to proactive behaviour; disincentives are the factors inciting an actor to reactive behaviour or even blockades.

Variety
Finally, the process architect sketches an image of the multiformity of the participating actors. Are actors homogenous or are there great internal differences? Does a party representative speak for this party or is this party so internally multiform that the representative only speaks for part of this party? A clear picture of the multiformity gives the process architect an insight into the stability of an actor's points of view. This is because an actor whose composition is highly multiform may be less stable in its points of view than an actor with a uniform structure.

An actor scan is a continuous activity
A remark about the actor scan should be made here. For many organizations, there is nothing new about an actor scan. Organizations realizing their dependence on other actors to achieve their aims also realize that such a scan is indispensable for making progress. They will perform such a scan (a field-of-force analysis). This produces a list of actors, their views and their resources, telling the initiator which actors to work on if he is to score a result.
Such field-of-force analyses tend to be once-only actions, performed when the initiator becomes aware of his dependencies. The dominant aspect of these field-of-force analyses is that the initiator wishes to realize something and that other actors have to be induced to cooperate or at any rate not to use their blocking power. From this perspective, actors are barriers that have to be overcome and the field-of-force analysis is a once-only means of achieving it.

The actor scan as part of a process design is something far less instrumental. Rather than being a once-only activity, the actor scan should be permanent, embedded in the process. The reason is twofold: interests are not fully knowable and interests as well as views may change in the course of the process, for example under the influence of the process itself.
Unlike the field-of-force analysis, the continuous actor scan aims to disclose views, interests and incentives allowing a process to be organized that results in an outcome recognizing the maximum number of interests. Actors actively participate in the process for this purpose.
One of the consequences of this active participation is that actors experience learning processes. They become aware of other actors' views, participate in research processes, communicate intensively with other actors in the process, et cetera. They will be influenced by this information and these networks. Their views will develop and they will learn what they consider to be their actual interests. In short, actors participating intensively in a process will further develop their views and interests. The consequence for the actor scan is that its outcome will be dated after some time. In other words, an actor scan requires updating, also because of the process itself. An actor scan that lasts for a long time may be a sign that a process is not developing properly.

4.4.2 Quick scan configurations [42]

The information gathered may enable the process manager to make a quick scan of the configuration of the parties. The architect takes the initiator's issue and then lists the views and interaction patterns of the actors involved.
The analysis may reveal which actors hold relatively extreme views and which actors propagate views finding broad support. It also becomes clear which actors communicate frequently and intensively and which actors operate in social isolation (see figure 4.2).

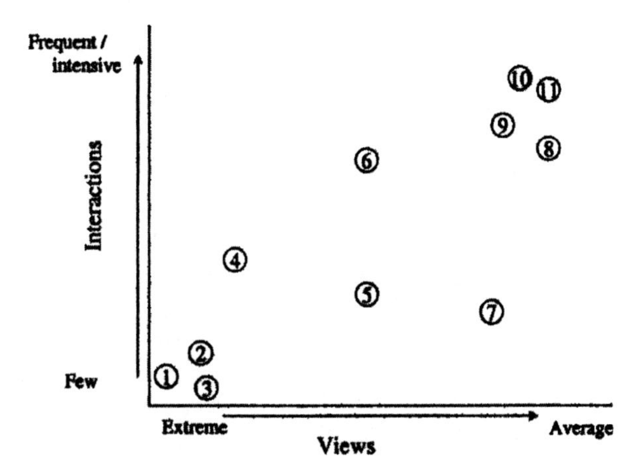

Figure 4.2 The position of actors in a network

Actors 1, 2 and 3 are marginal. They hold extreme views, which may even differ sharply between themselves, and they do not communicate frequently. Actors 8, 9, 10 and 11 occupy a central position in the network. They hold views close to the average view and communicate frequently and intensively.

In the first place, this analysis shows which actors occupy a key position in the network from a substantive and communicative perspective. They are the actors that have a well-developed network of connections and hold views that come close to those held by others. For actors of this kind, cooperation in a process tends to be self-evident: process agreements tend to codify behaviour of a kind that is natural to them anyway.

The analysis also shows which actors are marginal. These are actors holding an extreme view besides maintaining relatively few relations.

The analysis may be expanded by inventorying the distribution of resources over actors. If the actors in the top right of the figure (8, 9, 10, 11) are at the same time the actors with many resources, the process will take its course. The actors with intensive communication and average, widely supported views also have the resources in this case. They may therefore be expected to communicate properly about the use of their resources and to use them in a direction enjoying wide support. The situation is more difficult if the resources are concentrated among actors in a marginal position. It is very important for the process that they should remain committed to it, but their marginal position shows that this is not self-evident. Ways of committing them all the same are:

- reframing and renaming the process, allowing them to identify more closely with the process and thus encouraging them to communicate more intensively;
- coupling of the process to other processes and issues, in which they hold more average views and also maintain more relations.

The actor scan results in a list of problems featuring in the process. The process designer draws up a list of problems, distinguishing substantive and process-based problems, which we will deal with below.

4.4.3 Scan of the substantive couplings and the first agenda

While scanning the actors and analysing the configuration, the process architect has formed an image of the issues that make a process attractive to the parties. This produces a picture of the substantive couplings between issues that may be made during the process. The process architect formulates the agenda that forms the start of the process, based on the scan of the substantive couplings.

Converting issues and interests into dilemmas and 'dilemma sharing'

When drawing up the agenda, the process architect is faced with various substantive views of the parties, many of which will be conflicting. Opposite views are a source of conflict. How should the process architect deal with them? The answer is: by formulating them as a dilemma where possible. We define a dilemma as a problem the solution of which brings advantages as well as disadvantages, which are each others' counterparts.[43] There is not one single, obvious, exclusive solution, but there are several, each with its strengths and weaknesses. Framing a conflict as a dilemma has a number of positive effects on the process.

- Formulating conflicting views as a dilemma releases the process architect from having to make substantive choices; the parties involved will find that the process architect recognizes their views.
- A conflict will be reduced by the parties acknowledging that a difference is a dilemma. After all, if conflicting views are a dilemma, the question is no longer which view is preferable, but how the two views can be reconciled. This forces parties to think about a trade-off between the different views.
- Any dilemma will automatically prompt the parties involved to put their own views into perspective, since a dilemma implies that both views, though conflicting, may be correct.
- The amenability to process agreements increases if the parties are able to put their views into perspective, since such agreements aim to reach a trade-off between the extremes of a dilemma in mutual consultation.

Evidently, the dilemmas can only fill this function if the parties can sufficiently identify with them. Consequently, contriving dilemmas is an activity the process architect should share with the parties in the process. 'Dilemma sharing' is a crucial activity in making a process design. In many cases, a party goes through a learning process here: it first formulates its view, learns that it cannot be realized without the support of other parties, and is then prepared to participate in a process. If this party is subsequently prepared to frame its view as part of a dilemma, it indicates that other points of view are both defensible and reasonable from the perspective and the interest of another party. In fact, it thus shows it is prepared to participate in the process. The process of contriving and sharing dilemmas during the design of the process agreements is thus an important breeding ground for successful process management.

4.4.4 Substantive dilemmas and fixing the agenda

The process architect orders the substantive dilemmas in relation to each other, indicating how the various dilemmas relate to each other from a substantive perspective. For example, the choice of a particular dilemma may strongly influence the options in another dilemma.
The ordering may consist of a kind of decision tree, but the dilemmas may instead be incorporated in a number of clusters, giving them the outlines of a package deal.

The process architect also proposes the order in which the dilemmas in the process will be dealt with. He then decides about the substance of the agenda.
When fixing the agenda, the process architect should estimate the way he will be able to deal with each of the dilemmas. Some will be easy to solve, others will be difficult to solve. Then, drawing up the agenda, he may try to order the dilemmas in such a way that they promote a good process. This ordering may imply his spreading the discussion of the dilemmas intelligently over a certain period or coupling particular dilemmas and decoupling others.
A number of strategies a process architect can use to deal with dilemmas are given below. A number of these strategies will either remove or counteract the dilemma: a choice has either become unnecessary or may be postponed.

- *Resolving the dilemma by synthesis* The process architect foresees the possibility that the parties will fully agree with each other. The solution is new and fully meets the parties' interests. He therefore reserves time in the process to have the actors discuss the views that form the poles of the dilemma with each other and, if necessary, with third parties. A well-known example: two parties both claim a consignment of oranges. The conflict seems insoluble, as both parties claim exactly the same and what one of them receives will be deducted from the part the other party acquires. At some stage, the discussion shifts from the points of view to the two parties' interests. The one party's interest, it turns out, is the continuity of its soft-drink factory. It needs the oranges for the orange juice. The other party has an entirely different interest. It makes perfumes and needs the orange peels for them. Now that the two parties know each other's interests, they are able to get rid of the conflict in a most elegant way. The solution is obvious in this case. One party receives the flesh of all the oranges, the other party receiving all the peels. Both parties are completely satisfied, because their interests have been taken into account. Had the discussion not been conducted at interest level, the parties might have decided on a compromise of 50% of the oranges for the one and 50% for the other. The parties have reached a much finer outcome by negotiating on the level of interests.

- *Developing options in tandem* The parties agree to examine whether it is possible to implement both proposals simultaneously and in tandem. The parties examine until when this will be possible and what conditions must be met to develop the two proposals in parallel. The parties agree a monitoring process that will produce data to discuss the dilemma again later, but then on the basis of more information than is available now and jointly gathered information.
- *Pilot option* The parties decide to fully implement one of the options and to introduce the other option to a limited extent, as a pilot. The actors organize a process in which they jointly monitor the results of the pilot. The parties will reconsider their earlier decision if the pilot produces positive results.
- *Mothball variant* The parties choose one of the options and decide to keep the other option available as much as possible, too. This implies that resources are allocated to maintain the knowledge about this option, that–if relevant–the option is not made definitively impossible as regards room for it, et cetera. If required, the variant rejected in the decision making can be taken out of mothballs at once. In the decision-making process addressing the question whether Amsterdam Airport Schiphol should expand or whether an airport should be built in the sea, the outcome was that the North Sea option was mothballed. The parties decided to choose the Schiphol option in principle and maintain the knowledge necessary to develop the North Sea option should it ever be required.
- *Growth model* The parties explore the possibility of keeping open the option whereby the one option develops into the other option in the course of the time. In the discussion about expanding the Maasvlakte area, several spatial models were debated, including a large expansion and a smaller one. In such a situation, a growth model is an interesting option. Parties then choose the small Maasvlakte, while keeping open the option to develop the large variant at a later time, spatially, financially as well as administratively.
- *Removing the sting from the conflict by addressing the underlying question* The process architect examines whether there are any factors behind the dilemma on

which the survival of the conflict depends. If he traces them, he designs a research or advice process that should lead to broadly supported conclusions about the validity and relevance of these factors. In a discussion about the question whether non-disposable packages or disposable packages are better for the environment, such an underlying variable was found to dominate the discussion. The underlying factor turned out to be the number of times a non-disposable package is used ('the number of trips'). Most of the parties that were so sharply divided about the question whether disposable packages or non-disposable packages are better for the environment were found to endorse the following proposition: "the non-disposable bottle is good if many trips are made, the disposable package is good if few trips are made". The sting of the conflict appeared to be in the number of trips made by packages. Some parties thought that the average non-disposable package made many trips before becoming worthless, other parties did not believe that and argued that non-disposable packages only made few trips. The process architect detecting such an underlying factor may design a research protocol to solve the dilemma. The surprising outcome for all parties was that this sensitivity is very limited. The number of trips proved far less decisive for the difference in environmental effects than actors believed at the start of the process. This removed the sting from the conflict.

- *Sensitive pros or cons in the dilemma* The process architect examines whether the conflict concentrates on a few pros and cons of the poles in the dilemma. If so, the parties agree about the sense of the distinction between the two poles in the dilemma. They also agree on a valuation of most of the pros and cons of the two poles, but they differ so sharply about some pros and cons that the whole conflict centres on them. In such cases, the process architect may propose a research protocol that throws more light on these sensitive pros and cons.
- *Designing mitigating and/or compensating measures* If the effects of particular options are inevitable and parties largely agree about their negative consequences for other parties, the process architect may outline a process that results in measures mitigating or compensating the negative effects.

A process design results from negotiation (section 4.2). Laborious though this may seem, the above description of activities shows that the boundary between designing and managing processes is not always distinct. Parties reformulating views as dilemmas and negotiating about how these dilemmas should be addressed not only design a process, but have already partially started the substantive negotiations.

4.4.5 Process dilemmas and fixing the rules of the game

The process architect solves as many of the process dilemmas as possible so as to establish process rules that take the two poles of the dilemma into account.

Experience shows that almost all process dilemmas are amenable to this. Examples of common process dilemmas are:

Carefulness versus speed.
Should the process primarily be a fast one, taking the risk of some carelessness, or should it be organized as an extremely careful one, even though this takes a lot of time?

> The process arrangement uniting both poles may read as follows: "Actors shall compare alternative options submitted to solve a problem by means of a quick scan. If there is any doubt after the quick scan about the validity of the results, the parties may request a full, detailed analysis."

Many or few parties
Should many actors be involved in the process or should the number of actors remain limited?

> The participating parties may all be given the same roles within the process. Instead, differentiation is possible: there will then be parties participating in the decision making, for example, and parties that are heard or informed before any decisions are taken. The corresponding process arrangement may be: "The process shall distinguish an inner circle and an outer circle of actors. Particular matters shall only be dealt with in the inner circle. The actors in the outer circle shall be asked to advise on the matter. If one of the actors in the outer circle announces that it has a major interest in a decision, it shall count as a member of the inner circle in respect of that one decision. "

Confidential versus public
Should research findings and the substance of advice generated in the course of the process remain confidential or are they made public?

> The process arrangement may imply that everything is public, unless one of the participants wants to keep something confidential, stating the reasons.

Once the problem and the dilemmas have been formulated, most of the sensitive issues have been described. Negotiations will then be held about the rules. Agreements will have to be made about the following issues:

- *Entry and exit rules* The entry and exit rules describe which parties will participate, under what conditions parties are allowed to join the processes and how they can exit the process.

- *Decision-making rules* These rules stipulate how the parties reach decisions: when there is consensus, for example, or by majority decision making, in which case rules may then be agreed about how to deal with losing minorities. The decision-making rules tend to include agreements about conflict handling. Conflicts may be settled by voting, by passing on the conflict to another body, by arbitration, et cetera.

- *Organic rules* These are rules stipulating the organization of the process. The process architect describes the bodies required: a steering committee, for example, work groups and a group monitoring quality and progress. In some cases, these groups may require some kind of standing rules. Agreements also have to be made about the chairmanship and the secretariat of these groups. Last but not least, agreements are needed about the role of the process manager: what is the process manager's profile and what role does he play in the decision-making process?

- *Rules about planning and budget* The plan describes what activities will be performed in the process in what order with what deadlines. It also contains an estimate of the cost of the activities and the process management and describes who will bear what costs.

The concrete formulation of these rules will be guided by the design principles from chapter 3.

4.4.6 Testing the process design

The architect will test his first process design. He can do so in two ways. He either does it by himself or in a small group (prima facie) or he develops a simulation in which the players set to work with the process rules designed.

Prima facie test
The architect organizes a brainstorming session about the designed rules. The participants in the session consider the process critically, the key question being whether the process design serves their interests and is likely to be successful. Does the process match the incentives and disincentives established earlier for each actor? Have actors got options for behaviour outside the process that may be interesting for them? If so, are there any guarantees that the actors will remain committed to the process? Is there sufficient pressure on the process? Is the sense of urgency sufficient to commit actors to the process?

Simulation
More detailed test results will become available through a simulation game. The architect develops a simulation game. The game imitates the reality of the

decision making as much as possible. The players in the game then set to work, observing the process rules. Watching their behaviour, the architect analyses the functioning of the process rules and evaluates the results. If required, he may adapt rules on the basis of the results.

4.4.7 Participants

The last activity is finding the participants in the process. Of course, this is a crucial moment because much of the process depends on the behaviour of individuals. A number of questions should be asked when finding the participants.

Involved or not involved in designing the process?
As we observed earlier, designing a process is an important learning process for all parties. That is why it may be desirable for the intended participants in the designed process to have been involved in it. Instead, it may be desirable to appoint other persons as representative in the process, for example, if there were many conflicts during the design of the process.

Mutual right of consent?
Such a right implies that a person can only join a process on behalf of a party if the other parties give their consent. One advantage of this may be that there is less risk of 'incompatibilité des humeurs'. It also reduces the risk of asymmetrical representations.

> The organizations participating in the process have different interests in the process. Some parties are passionate: they have great interests in a successful outcome of the process; other parties are less passionate. This may have important consequences for the representatives sent by the parties. There is a risk that the passionate party will send a heavyweight representative, for example from the top level of the organization, while the other party sends an intermediate-level representative. Such asymmetry may disturb the process.

One disadvantage of this method is, of course, that it enables parties to meddle with the internal affairs of other parties or make a strategic use of this arrangement.

Direct or indirect representation?
Direct representation means that a party appoints a representative. Indirect representation means that somebody participates in a process on behalf of a party, but does not act as its formal representative. The indirect form of representation may be used if parties think it important that a particular party

should participate in the process, but this party cannot be persuaded to do so. This indirect representative will then have to 'earn' his 'own' party's commitment during the process.

If a process has been designed, however, and the parties have committed themselves to this process design, the process manager will enter the picture. With the help of the process rules, he should induce the actors involved to take decisions.

4.5 PROCESS MANAGEMENT AND RELATED APPROACHES

In this section, we will discuss the positioning of process management as compared with related approaches of decision making.
How do our definition and approach of process design and process management relate to procedural rationality, consensus building and interactive decision making?
Our discussion does not aim to give an exhaustive description of these approaches; we only compare the related approaches in order to position process design and process management more precisely.

4.5.1 Procedural rationality

In his classic article "From Substantive to Procedural Rationality", Herbert Simon introduces the distinction between substantive rationality and procedural rationality.[44] From a substantive perspective, behaviour is rational if it contributes to the realization of given objectives. According to Simon, behaviour is rational from a procedural perspective if the process, resulting in a decision, is 'correct'. One of Simon's definitions of 'correct procedures' is adequate thinking processes and the use of the right algorithms.
The similarity between process management and procedural rationality is obvious. Both approaches accept that it is impossible in many cases to design the best substantive solution for a substantive problem. This is why both approaches choose not to seek substantive solutions only for problems any longer, but also to pay attention to the correct process, which should then result in good solutions from a substantive point of view . Both approaches assume that a well-designed process will result in a good solution as regards substance.
There are also important differences, however, between the process approach and the idea of procedural rationality.

Process as a rational design or as a result of negotiation
First, according to Simon, the correct process is again a matter of rational design. In his view, there is a best procedure to approach a particular problem. He

therefore speaks of *procedural rationality*. Consequently, he argues, finding and designing the correct procedure is primarily an academic activity. Once the best procedure has been designed, it may be prescribed. Where appropriate, problems should be solved by means of that process.
In the process approach, the process design is the result of negotiation. The parties involved commit themselves to a process design because they feel the process is fair and offers them enough opportunities of promoting their interests. Every actor decides for himself in how far this is the case for a concrete draft process design. The fact that every actor forms his own judgement about the quality of the process is difficult to match with the idea of a process that is the best from an objective point of view thanks to a scientific design. A process is good because it is acceptable and confidence-inspiring to the chief stakeholders. In other words, process management does not lead to a process designed once, applicable universally and perpetually in addressing problems of a particular type.

This does not mean that scientific reflection on process rules is impossible. Of course, accumulation of knowledge about process rules and how they operate is possible. Of course, professionals can build up knowledge about the effectiveness of particular process rules and use it in new cases to advise about how to design processes: the design principles we have formulated are an example. However, these principles are merely input for an interaction process between the process architect and the parties involved.

Second, parties that have to solve problems in a network will *learn*. If a particular party tries to influence another by using a management tool, the other party will learn how to deal with this instrument and also how to evade its steering effect. In the context of a network, the effectiveness curve of an instrument will look as follows:

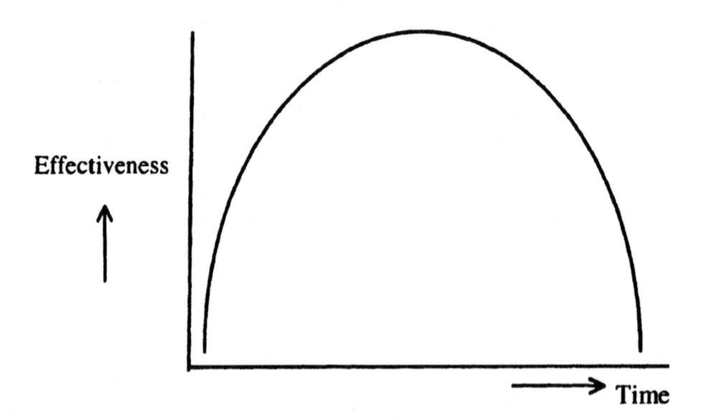

Figure 4.3: Effectiveness of tools in a network

The more the same process design is used, the more the parties involved will learn how to evade its steering effect: for example, they learn what the weaknesses of a process design are, which they then utilize to advance their interests. This may cause the process to lose its value for other actors. This means that the effectiveness and usefulness of the same process design will decline after some time. This Law of Decreasing Effectiveness implies that there is no 'best' process that will continue to be the best. A process design will cease to be effective at some stage.

4.5.2 Consensus building

Of course, process management is closely related to consensus building, which may be defined as 'a process of seeking unanimous agreement. It involves a good-faith effort to meet the interests of all stakeholders'.[45] One of the key techniques of consensus building is to abstract stances adopted by parties and direct the attention at parties' underlying interests. The 'manager' has a facilitating and mediating role. Examples of 'facilitating' are ways of having groups communicate effectively face-to-face. Mediation aims to bring parties' sharply different views closer together. Mediation may take a rather longer period, aims to prevent a 'zero sum game' and direct the process to new solutions, producing added value for all parties.
Such consensus-building methods may be part of process management, but there are important differences as well.

Process management is embedded both institutionally and administratively
Consensus building is presented as a universally applicable method. It comprises a large number of steps to be taken in order to reach consensus. In many cases–not all, however–the focus is on interaction between persons.

Process management, on the other hand, is embedded in existing administrative processes. All sorts of phenomena occur in these processes, which form part of the administrative game.

- Parties practice power play when it suits them. They use their position of power to press home their views; consensus may also evolve under duress (also see chapter 7).
- Parties know that they will meet again later to discuss another issue, and may exchange the results of the negotiation in the current process for future results in the next issue.
- Parties play strategic games: they misrepresent matters, play the waiting game, above all want to keep their options open, have no interest in consensus, et cetera.

The fact that process management is embedded in an administrative process, means that it should be able to cope with such and similar behaviour.
This largely means that process management enables parties to show their natural behaviour (also see the intermezzo in chapter 8). A party choosing a strategy in which all options are kept open is given room for it; there is no such thing as an obligation to state preferences. Nor does it make any sense to lay down the rule that parties have to be open to each other during the process or study each other's interests. It is wiser to allow parties room for their natural behaviour.

Administrative embedment also implies that the process architect is sensitive to the risk that a process will be terminated if parties show their natural behaviour. For example, joint 'fact finding', an important element of consensus building[46], may pervert because all parties have an interest in constructing an incorrect image of facts and causalities. Administratively, it is naive to suppose that parties will always cooperate honestly in establishing facts and causalities. This is why a process design should contain arrangements to reduce this risk (see section 3.6 about the role of experts).
It should be pointed out that embedding process management in administrative processes also implies that there is no universal process design. A process design will differ in accordance with the specific administrative context.

Related methods	**Similarity to process management; both ...**	**Difference with process management; process management ...**
Procedural rationality	... postulate that only a process can be designed, not its substance.	... has a process resulting from negotiation; ... has no 'long-life' processes.
Consensus building	... concentrate on interests and avoiding ' zero sum games'.	... is administratively embedded and tolerates administrative behaviour.
Interactive decision making	... are open as well as contingent.	... is administratively oriented and less non-committal.

Table 4.1: Process management compared with related approaches

4.5.3 Interactive decision making

In interactive decision making, the body competent to take a decision involves other actors, such as citizens, firms and interest groups, in taking the decision.[47] Interactive decision making has boomed in developing spatial plans, such as

zoning plans, city plans and plans for the construction of infrastructural projects. Interactive decision making builds on the tradition of participation in developing spatial policy.
Interactive decision making has a number of features in common with process management. The first similarity is that the actor initiating the interactive decision making evidently wishes to admit other actors to his decision-making process. Process management has similar features. It also has an actor who has an interest, who wants decisions to be taken about it and who wants to exchange opinions with other actors. The second similarity is that, like process management, interactive decision making is highly contingent. It assumes a shape depending on the specific questions asked and the nature of the actors whose interests are at stake.

Process management is administratively oriented
There are a number of differences between process management and interactive decision making, however.

In interactive decision-making practice, it is usually a public body that develops a plan in cooperation with citizens, firms and societal organizations concerned. Process management is oriented towards actors, cooperating in administrative or managerial processes. In other words, process management has strong administrative or managerial properties, while interactive decision making has a societal orientation.

Interactive decision making tends to start with a government that has to take a decision or establish a plan. This is why designing the interactive process also concerns this government: the government lays down the rules for the process, defines the subject of the process, decides when the interactive process will start and during what period the interactive decision making will take place. In process management, establishing the process is an activity in which more parties are involved.

The latter is due to the fact that interactive processes hardly ever concern envisaged decisions. They usually concern plans, views or policies. If they concern decision making at all, the relation between the processes and the prospective decisions is a very indirect one. Interactive decision making therefore risks degenerating into a non-committal activity.
Managerially intelligent actors who know this, also know that participation in interactive 'decision making' has limited significance. They will either participate carelessly or not participate at all. Interactive decision-making processes tend to have a moral connotation: from a normative perspective, it is desirable to involve parties in the 'decision making'. Process management has

less of this moral connotation; it is rather an approach that may be helpful when decisions have to be taken in a network and parties know that they are interdependent. Parties will only join a process if it has something to offer them; this implies that, in most cases, the process should involve decisions to be taken. Since process management involves decisions to be taken by administrative parties, there is great attention for the natural strategic behaviour of these parties (also see above). Interactive decision making tends to differ in composition: for example, the parties have to adopt an open attitude and be receptive to each other's interests. A significant detail is that, in many discussions of interactive decision making, a great deal of attention is paid to the use of proper communication techniques. As we pointed out earlier, it is unrealistic to expect managerial parties that have to take a decision affecting their interests to join such a process automatically.

Notes

[40] This and the following chapter are partly based on process designs and outlines of process designs we made. J.A. de Bruijn and E.F. ten Heuvelhof, *A process design for a study into travel behaviour*: Memorandum of advice commissioned by the Ministry of Education, Dutch Rail passengers, regional transport and VSV+, TPM, Delft, 2000 (in Dutch); J.A. de Bruijn and E.F. ten Heuvelhof, *Rules of the game for the negotiations between the Ministry of Education and the public transport companies about a new season ticket for students*, Delft, 2000 (in Dutch).; J. A. de Bruijn, L. Geut, M. B. Kort et al.: *Procedures for decision-making about a national airport*. The Hague, 1999. Report commissioned by the Ministry of Transport (in Dutch).; J.A. de Bruijn and E.F. ten Heuvelhof *A process approach for a product-based environmental policy,* Delft, 1999 report commissioned by VNO-NCW (in Dutch).; J.A. de Bruijn, R.J. in ' t Veld et al., *Procedures for the project management of Mainport Rotterdam*, The Hague, 1999 (in Dutch).; J.A. de Bruijn, E.F. ten Heuvelhof and H.I.M. de Vlaam, "Interconnection Disputes and the Role of the Government between Substances and Process, in: *Communications & Strategies* (1999), pp. 295-317; J.A. de Bruijn and E.F. ten Heuvelhof, *A quality system for policy and policy processes*, Delft, 1999. Report commissioned by the Ministry of the Interior (in Dutch).; J.A. de Bruijn and E.F. ten Heuvelhof, *Package Covenant II: Clustering and Process of Reporting Protocol. Report for the Ministery of Housing, Environment and Spatial Planning and the Dutch Packaging Industry*, Delft, 1999; J.A. de Bruijn and E.F. ten Heuvelhof, *A design for a protocol for testing ALARA efforts*, 1997, commissioned by the Branded Goods Foundation (in Dutch); J.A. de Bruijn and E.F. ten Heuvelhof, *The implementation of the Malta Treaty: A design for process-based governance*, Delft, 1996 (in Dutch).; J.A. de Bruijn and E.F. ten Heuvelhof, *A Design for the Decision-Making Process about Maasvlakte II*, Delft, 1995 (in Dutch); R.J. in ' t Veld, J.A. de Bruijn, E.F. ten Heuvelhof et al., *Recommendations for a Process Standard concerning the Environmental and Feasibility Analysis as laid out in the Dutch Covenant for Packaging*, Rotterdam, 1992.

[41] John P. Kotter, "Leading Change: Why Transformation Efforts Fail", in: *Engineering Management Review* (1997).
[42] From: C.J.A.M. Termeer, *Dynamics and inertia surrounding manure policy*, The Hague, 1993 (in Dutch).
[43] M.J.W. van Twist, J. Edelenbos and M. van der Broek: "The courage to think in dilemmas", in: *Management en Organisatie* (1998), Volume 52, No. 5, pp. 7-23, 9, Alphen a/d Rijn (in Dutch).
[44] H.A. Simon, "From Substantive to Procedural Rationality", in: S. Latsis (ed), *Method and Appraisal in Economics*, Cambridge, pp. 129-148.
[45] Susskind, McKearnan and Thomas-Larner (1999), p. 7.
[46] J.R. Ehrmann and B.L. Stinson, "Joint Fact-Finding and the Use of Technical Experts", in: Susskind, McKearnan and Thomas-Larner (1999), pp. 375-401.
[47] E.H. Klijn and J.F.M. Koppenjan, "Interactive Decision Making and Representative Democracy: Institutional Collisions and Solutions", in: O. van Heffen et al., *Governance in Modern Society*, Kluwer, 2000, pp. 109-134; Edelenbos (2000).

PART III MANAGING THE PROCESS

5. THE PROCESS MANAGER AND THE OPENNESS OF DECISION MAKING

5.1 INTRODUCTION

This chapter examines how the process manager can ensure that the decision making is an open process: the relevant parties have to be involved in the decision making and they must be certain that their interests will be protected within the context of the process agreements where possible. This chapter therefore discusses the experiences with the three design principles relating to the openness of the decision making. Section 5.2 highlights the design principle that forms the start of the process approach: involving other parties in the decision making. A number of criteria that can be used here have been mentioned in the preceding chapters. A number of follow-up questions will be dealt with now. Tension exists between the need for openness and the need for control. How should it be dealt with? What should be the next step when it is not clear to an initiator what parties are needed because this only becomes clear during the process? What should be done when an initiator needs these parties, but they nevertheless refuse to participate in a process? Section 5.3 focuses on the transformation from substance to process. When can a process manager apply this principle and how should he prevent the risk of a process scuppering due to proceduralism? This section also discusses whether any substantive frameworks should be imposed on a process. Section 5.4 addresses the requirement that the process should be transparent. We discuss the question what functions a process manager can fulfil and present two perspectives of a process: an instrumental perspective and a perspective in which a process has an intrinsic value.

5.2 INVOLVING PARTIES IN THE DECISION MAKING

The criteria for involving parties in the decision making were already discussed in chapter 4. This section will present three issues with regard to this design principle: how to control a process; the unknowability of actors, interests and resources; and parties' refusal to participate.

5.2.1 Controlling a process

The openness of the agenda

The principle of openness implies that the parties to be involved also influence the process agenda. After all, they will only join the process if it is interesting for them; the process becomes interesting when subjects are discussed that affect the parties' interests.[48] As a consequence, the process agenda may be longer than the initiator intended originally (it becomes a 'multi-issue' agenda). In some cases, new items may be a reason to invite new parties because they have great productive power: the power to actually implement the new agenda items. The intuitive reaction to this tends to be that the process becomes too complex, will less easily produce results and that it is therefore imperative to reduce the agenda to a limited number of important items as soon as possible. From a project perspective, this is an attractive strategy: it facilitates control of the process. From a process perspective it is desirable, however, to preserve the multi-issue nature of the decision making as long as possible. There are two reasons for this:

- An agenda containing a great many items allows these items to be coupled. Party X has an issue that can be solved by coupling it to the issue of party Y because Y has the required means. Party Y is prepared to make these means available because Z no longer blocks an issue attractive to Y, etc.
- Moreover, an agenda with a great many items nearly always reduces conflicts. This is easy to explain: every issue will have a different coalition of proponents and opponents. If so, it is hardly attractive for one party to stir up a conflict with another party about a particular issue; after all, the other party may be a coalition partner for another issue, which imposes on the one party moderate behaviour towards the other party.

The tension between openness and the need for control

Although complexity (in the sense of many parties putting forward many items for the agenda) may therefore be conducive to decision making, there is, of course, a limit to the number of parties and issues dealt with in the process. In other words, there is tension between the openness of a process on the one hand and its controllability on the other hand. How to deal with this tension? Two types of strategies are possible.

The first strategy is that, ahead of the process, the initiator selects the parties to be involved and also forms an opinion about what items should be placed on the agenda. The advantage of such a course of action is that a controllable process is designed, but a major disadvantage is that the excluded parties cannot oppose the process. Nor is it easy in many cases to estimate before a process who the important parties are and what parties can be kept out. Something similar applies to the process agenda: what, at the start of the process, seems a detail that need not be placed on the agenda may become a main issue during the process. This is why this first strategy is effective only if the initiator can be absolutely sure that the limitations he imposes will be no point of discussion during the process.

A second strategy is that, at the start of the process, the initiator does not allow the value of control to play a part yet. He generously invites parties and also generously allows them to put forward items for the agenda. The underlying idea is that control should not be imposed on the parties beforehand, but that it will develop *during the process*. The parties will learn during the negotiations that the open nature of the process makes it insufficiently controllable. They conclude, for example, that the large number of items on the agenda hampers decision making. In such a situation there are two possibilities:

- A critical mass of parties has no interest in taking decisions and is satisfied with the stalemates that have arisen.
- A critical mass of parties is dissatisfied with the stalemates that arise and wants to make arrangements for enhancing the controllability of the process.

In the first situation, there is evidently no support for taking decisions. The process architect has selected parties and agenda items and has done so in consultation with the initiator. Evidently the initiator also concluded that the selected parties and items are necessary to serve his interests. The conclusion should then be that evidently there is insufficient support yet to arrive at quick decisions. In the second situation, there will clearly be support for proposals to improve the controllability of the process. These may imply, for example, that particular parties settle for a marginal role in the process or have this marginal role imposed on them by other parties. Particular agenda items may also be temporarily removed from the agenda, be reformulated or instead be dealt with quickly; or a number of roles may be distinguished in the process (taking decisions, listening, advising, etc.). It is important that such forms of control are accepted by the parties beforehand, because they evolved during the process and were suggested by the parties themselves (at least, by a critical mass of parties). In the hypothetical case that an excellent process manager works out exactly the same control measures at the start of the process, arrangements made during the process are always superior to preplanned arrangements. This is because what evolved in a process has the support of the process parties; what was worked out beforehand lacks such support. As soon as a preplanned arrangement creates any disadvantages, the parties will blame the process manager for them and he will be expected to solve the problem. If an arrangement made in the course of the process creates any disadvantages, the parties will have to solve this problem together. Moreover, there will always be parties disliking the role allotted to them. A division of roles will have more authority, however, if the parties as a whole are responsible for it, than if the process manager or – which is worse – the initiator is responsible for it. Parties opposing the role allotted to them will have conflicts with all other parties in the first case and only with the process manager or the initiator in the second case.

5.2.2 The unknowability of actors, interests and resources

It has been assumed so far that the process architect and the process manager are able to form a picture of the dependencies in a network and of the actors to be involved. It may be unclear, however, what actors are to be involved in the decision making, what their interests and resources are. It may likewise be unclear what actors are necessary to enrich the decision making.

The uncertainty about interests has another dimension: it may be found during the process that the process *architect* has misjudged the relevant actors, their interests and their resources. Actors that ought to play a role because of their interests and/or power may not have been admitted to the process.

This may jeopardize the process. Adjusting the process (involving new actors, making new agreements) may add to the chaos. Failing to adjust the process may mean that new actors feel insufficiently represented in the process, to the surprise of the process architect in some cases.

> This is illustrated by the course of events in the negotiating process between industry and societal organizations about packages referred to earlier. Long after the start of the process, a request is received from the industry for permission to appoint an extra representative in the steering committee for the non-food packers/fillers (the seat for the packers/fillers is filled by the food sector). The underlying idea is that the interests of the non-food sector cannot be looked after sufficiently by a representative from the food sector. The request is turned down, because it would prejudice the balance in the steering committee. In principle, this is a risky development: an important party finds itself underrepresented in the process. This may affect the open nature of the decision making. The refusal to include the non-food representative has not prejudiced the process, however. One major reason for this is that the industry representative had organized consultations with the group he represented very systematically. Consultations with representatives at sector or corporate level in the 'round table group' were held before each meeting of the steering committee. The group was accessible to any company involved. This gave the companies a permanent account in the process. Something similar was true of the representatives of the Consumers Association and the environmental organizations. Although the groups they represented were not organized, they nevertheless formed the account for consumers and environmental interests respectively. [49]

The problem of the unknowability of the main actors, their interests and their resources may therefore be solved by:

- having the process architect formulate the interests involved at a high level of abstraction ('environment', 'industry', 'consumers');
- giving the representatives in the process the role of account: any views held within the wide interest they represent can be put forward through them;

- giving the representatives in the process no detailed mandate beforehand; this is counterintuitive, because it would seem to invite a non-committal attitude, but the advantage of it is that representatives have some room if there is any change in the actors' involvement;
- ensuring a heavy representation, allowing the representatives also to play an authoritative role towards those they represent and to convince them (they are not only the 'messengers' of those represented, but can also play the role of 'missionaries' towards those represented);
- allowing the representatives in the process to 'earn' the mandate and the commitment of those represented during the process.

Table 5.1 lists the differences between an 'account' and a classic representing role.

Representative	**Account**
At party level	At interest level
Great deal of commitment power	Little commitment power
Clear mandate	No mandate or an unclear mandate
Staffing: usually of a lower calibre than the represented: the 'messenger'	Heavy staffing; able to play the role of 'missionary' towards the represented
Commitment by those represented laid down before to the process	Commitment by those represented results from the process, has to be earned
Able to cope with dynamic if and insofar as the represented permit this and/or it fits in with the mandate; may be problematic.	Able to cope with dynamic if and insofar as the broad-based description of the interest permits; this will nearly always be the case.
Given the strict mandate, expansion of the number of those represented (and thus of those who decide about the exercise of the mandate) will be a problem.	Given the loose mandate, expansion of the number of those represented (and thus of those who decide about the exercise of the mandate) will be no problem.

Table 5.1 Differences between a representative and an account

Here, the same mechanism operates as in the control issue. In a particular situation, it may be neither possible nor wise to have all interests represented before the process and give the representatives a mandate to commit those they represent. Proper representation and commitment should mainly evolve during the process.

5.2.3 Parties' refusal to participate

Participation at the start and at the end of a process may differ sharply. Particular actors may take a limited interest at the *start* of a process. The main reason is that it is insufficiently clear to these actors what direction the decision-making process will take. Participation takes them time (and costs them money), while it is insufficiently clear what product it may yield. Accordingly, there is insufficient incentive for particular actors to participate in the process, although there is plenty of opportunity for them to influence the decision making.
The picture is just the opposite at the *end* of the decision-making process. There will be a keen interest among particular actors in participating in the decision-making process at this late stage; this is because the products of the process are gradually becoming clearer. The main decisions have now been taken, however, causing the possibilities for influencing to be limited. Although there are ample opportunities to influence the decision making at the start, the participation rate is low, whereas it is high by the time the possibilities for influencing are limited.

> Drawing up an area traffic plan requires the involvement of many actors: municipalities, the province, the national government, companies and the public. In many cases it is unclear at the start of the process what the eventual traffic plan will imply. This vagueness is a reason for particular actors not to participate in the decision making or to participate only formally. When the plan assumes a concrete shape, particular actors notice that their interests are insufficiently served and they start interfering with the decision making at this late stage. This may jeopardize consensus between the participants, which may have been reached after laborious negotiations.

Of course, the decision-making process is put at risk if actors behave as described above: a low participation rate at the start of the process, a high participation rate at the end. Ideally, a decision-making process moves from variety (start) to selection (end) (see chapter 8). If participation is low at the start, this may mean that insufficient variety is being generated. At the end of the process, no authoritative selection can take place because actors suddenly wish to participate and generate a new variety.

The process manager should prevent this. First, this mechanism confirms the need of a *sense of urgency*; a process that is started too early tends to be ineffective.
This mechanism also confirms the need to make clear process agreements. This is because the risk of the mechanism becoming activated is the greater as there are fewer explicitly defined process agreements.

Finally, it is important for the process manager to convince the parties involved at the start of a process that it is attractive to participate also at that stage. Many writers point to the need for "distributing profit" to the parties also at the start of a process.[50] If the parties have to participate for a long time before having any prospect of gain, there is a major risk of start-end problems. The start of a process becomes attractive, too, if results can become manifest in the early stages of a process. In some cases it is recommendable for the well-motivated parties to make sure they already have something to offer to the other, less motivated parties in the early stages of a process.

5.3 THE TRANSFORMATION FROM SUBSTANCE TO PROCESS

An important design principle is that a minimum of substantive choices is made before the start of the process. All that is laid down is the *moments of choice* in the process, and it is subsequently indicated how the decision-making *process* will pass off at these moments. The idea is that the parties in the process take a substantive decision at the moments of choice defined beforehand, in accordance with the process agreements made beforehand.

This design principle can also play a role *during* the process. An intended substantive decision can be transformed into a process decision. This section deals with this mechanism. It describes under what circumstances such a transformation takes place, its risks, how the process manager can deal with these risks, and whether and how the process manager should indicate substantive frameworks.

5.3.1 Conditions for a transformation from substance to process

The transformation from substance to process implies that a procedure should be agreed when a substantive choice is to be made. For example, parties that have to make a choice between X and Y decide that they will first do research into alternatives, then add these alternatives to X and Y and subsequently take a decision. Such a transformation is necessary if parties are unable to take a joint decision. This will be the case in the following situations:

- when there is profound distrust between the parties;
- when there is uncertainty among the parties about the process or they have not got accustomed to the process yet;
- when there are a great many substantive uncertainties among the parties;
- when conflicting interests cannot be bridged, or cannot be bridged at present.

The proceduralization of decision making enhances confidence in the process and also allows parties to build up mutual trust.

If the parties nevertheless head for substantive decision making, there is a risk that a process will lose itself in an excess of detail and get bogged down at an early stage. This is because uncertainties and distrust (the lack of a joint frame of reference) will easily result in trenches between parties: every decision has to be secured. Every party will fear that a particular decision will restrict its own degrees of freedom or create a point of no return. As a result, the process manager should navigate between the Scylla of being blamed for proceduralism and the Charybdis of premature substantive decision making. In other words, he should ensure sufficient support for the process approach among the participants, even though it is impossible to take substantive decisions. Parties should learn how to recognize the advantages of procedural decisions. In the opening phase, a great deal of effort should be invested in internalizing the process approach. The following strategies can be pointed out here.

- *Every party benefits from the process agreements; the process manager makes the advantages of the process agreements clear to everyone.*
  In the first place, internalization can be achieved by making clear to the participants that they have an interest in or will benefit from the design principle 'substance becomes process'. There is every chance that party X will advocate substantive decision making if it feels (1) that a substantive decision serves its interests at that moment and (2) that it might find a majority to support a decision. Such a decision may come too early in the process because it may harm the interests of the other parties too early in the process. Consequently, at such a moment, party X will hardly support the taking of procedural decisions in accordance with the design principle 'substance becomes process'. Party X will have to conclude at some stage that other parties do find themselves in this comfortable position, which will prejudice party X's position. The process manager may have to make the participants aware that they can all benefit from the design principle. This may create support for the process design and give rise to some reserve towards premature substantive decision making.

- *Tolerance of irrelevant, but substantive decision making*
  An important mechanism in processes of this kind is that substantively interested parties have a pain limit, from where they no longer accept process management unless it is accompanied by substantive discussion and decision making. If this pain limit is exceeded, there is a serious risk that the process approach will be blamed for the shortcomings mentioned above. Accordingly, support can only be retained by ensuring a regular dose of substance: a substantive discussion, followed by decision making. In a particular situation, it may be clear to the process manager that such decision making will be reversed in the course of the process. For the sake of support for the process,

some degree of tolerance for substantive discussions not directly related to the final process result may be advisable. In such a situation, too, a process manager should not wish to control too many things, but allow the process to take its course.

5.3.2 Frameworks and crystallization points

A proper process offers room to the parties involved to define problems and find solutions together. An important question here is whether and how substantive preconditions can be imposed on such a process. An essential dilemma presents itself here. On the one hand, it is unthinkable that there should be no such preconditions; there must be a substantive reason for a process. On the other hand, pre-imposed substantive preconditions may frustrate the process if they offer the parties insufficient room, thus precluding learning processes.

> The latter disadvantage is even more serious if the following mechanism is considered. Processes only stand a chance of success if parties are offered room to define problems and solutions jointly. This poses a risk to some parties: they have a clear preference for a particular problem definition and problem solution, not knowing whether these will stand up in the process. Nevertheless, these parties will have to participate in a process because they can only thus gain the support of other parties. For these parties, an attractive way of dealing with this tension is to formulate preconditions beforehand. This limits the scope for the other parties, which improves the chance of an outcome sufficiently attractive to these parties. As a result, there will be a strong incentive for particular parties to impose preconditions beforehand if the process approach for change is started.

The question here is how substantive conditions can be attached to a process without constituting constraints. For this purpose, we will introduce the distinction between frameworks and crystallization points and revert to this question in chapter 6.

A *framework* offers room within predefined limits. Those who recognize that ex ante substantive planning of an activity is impossible should offer room to others. This room cannot be unlimited, however, which is why the extent of this room is defined ex ante by formulating frameworks. So there is room for innovation and creativity, but within predefined frameworks. Frameworks, however, can all too easily develop into rigid preconditions, which may be perceived as impediments and frustrate innovation.

Frameworks	Crystallization points
Development *within*....	Development *from* ....
Substance constrains	Substance challenges
Incentives for the absence of the party establishing frameworks	Incentives for the presence of the party establishing crystallization points

Table 5.2 Frameworks and crystallization points

A *crystallization point* is a substantive idea that can be worked out. A process set up round a number of crystallization points invites parties to discuss these crystallization points, to criticize and enrich them, and to couple them to other crystallization points. Crystallization points differ from frameworks in three respects.

- A framework should ensure that a development stays *within* certain substantive limits. Crystallization points aim to develop certain substantive notions. A framework is binding and fixes limits: the process has to move within the framework. Crystallization points offer possibilities: in the process, the crystallization points are developed further in the direction decided by the parties.
- A party with a highly substantive orientation, translating this orientation into frameworks, will hamper the other parties. If the same party translates its substantive orientation into crystallization points, it will encourage other parties. The richer the substance of a crystallization point, the stronger the stimulus it provides to the other parties.
- The third difference – which, however, need not present itself – is that a party establishing frameworks need not necessarily be active in the process. After all, the frameworks offer sufficient guarantee that the outcome will stay within the bandwidth acceptable to this party. For a party formulating crystallization points there is a strong incentive to participate in the process, however, since it will wish to be involved in elaborating these points.

A family that has to decide about its holiday destination can do so in four ways.

- The first strategy is substantive: the parents decide that the holiday will be spent in Sardinia from 1 to 21 August and that they will travel by air.
- The second strategy is a process; the parents indicate that a decision about spending the time between 1 and 21 August will be taken in a consultation process in accordance with particular rules of the game.
- The third strategy is also a process, but the parents establish frameworks: the holiday has to be spent in Italy, the family will use air travel and no destination will be chosen that the family has visited before.

- The fourth strategy also involves a process, but is shaped round crystallization points: an island, sunshine and culture, for example. Given these three crystallization points, an interaction process will develop, in which the members of the family consider a number of options in consultation with each other. This interaction process may take a surprising turn. It may end with the conclusion that the holiday will be spent in the Norwegian cold. One member of the family proposes Sicily, because of the wild mountains in the north of the island. In the following interaction process, the family conclude that this appeals to them: mountains by the sea. However, the dazzling sunshine in the deep south of Europe would be a problem to another member of the family; the search goes on for a combination of coast and mountains. Northern Norway suddenly becomes an option then; the fact that this fails to meet all three crystallization points is no problem as long as the parties involved are enthusiastic about it. After all, crystallization points are not frameworks.

Of course, the fourth approach does not mean that anything is possible. There will be budget constraints on the choice of a holiday destination. In the first approach, the parents have already taken this into account. They can fix the overall cost of the holiday because they unilaterally choose the destination and the mode of travel. In the third approach, they can lay down the budgetary preconditions in the framework: the holiday should not cost more than an x amount of money. This is different in the second and fourth approaches. In the consultation process, the parents will indicate that they have an x amount available to fund the holiday. Since they provide the funds, a precondition *develops during the process,* but it need not be a rigid one. Conceivably, it may eventually be decided that the family will not take a winter sports holiday in the year in question and that the budget thus made available will be spent on the summer holiday, allowing the choice of another destination. The latter shows that, if there are no frameworks, there is a strong incentive for the initiating party–the parents–to participate in the process. The cost of the holiday might turn out very high without their participation.

5.4 PROCESS AND PROCESS MANAGEMENT ARE MARKED BY TRANSPARENCY AND OPENNESS

An important building block for the process approach is transparency: designing a transparent and fair process makes it attractive to parties to participate in the process. Transparency means that they can check whether the process is sound and offers them sufficient opportunities to promote their interests.

Transparency may also be required with regard to the process manager's actions. An important notion in the theoretical discussions about resources tends to be that the process manager should play an independent, disinterested role in the decision-making process and derive much of his authority from it.[51]

It may also be concluded that the process manager's position and role are never a matter of course. To carry out his task, the process manager needs the parties' support. This creates a very delicate relationship between the process manager and the parties. He must be transparent and adopt an independent position between the parties, while depending on these parties to perform his task. Particular parties may have an interest in questioning this independence or may only be prepared to support the process manager on certain conditions.
This section deals with this relationship between the process manager and the process parties as follows: the different roles a process manager can assume and the risks attached to them; the function of progress of the process; and the role of the process agreements: are they merely instrumental or do they also have an intrinsic value?

5.4.1 The roles of the process manager: independent and dependent

The idea that a process manager can have interests would seem contradictory: after all, the process manager is the disinterested facilitator in the process. He occupies an independent position towards the other parties. In reality, the relationship between the process manager and the other parties is more paradoxical. The process manager is independent, but also dependent on the process parties. He will be vulnerable when support for his authority crumbles among one or more parties.
The first strategy to deal with this is to reinforce the process manager's position of power by giving him additional functions. The literature makes a number of suggestions here:

- making a substantive contribution: the process manager is also an expert on substance;
- controlling financial resources: the process manager is also the treasurer;
- keeping the balance: the process manager as 'countervailing power', as 'balancer', embodying particular underrepresented interests in the process.[52]

These roles are attractive in a *cooperative* environment. The process manager can make a substantive contribution, can exercise some control by the use of money and can maintain the balance in the process in his role of 'balancer'. An accumulation of functions will also make the process parties more dependent on the process manager, which may reduce his vulnerable position. An accumulation of roles will therefore enhance the problem-solving power of the process manager.
In a *non-cooperative* environment, these roles may turn against the process manager:

- substantive statements may be regarded as a choice in favour of one of the parties;
- the process manager may be blamed for any financial problems;

- a 'balancer' may be blamed for opportunism or partiality;
- the roles may conflict in an unfortunate way: a substance expert's stance may differ from a 'balancer's' stance.

If a process manager is nevertheless given such roles, the well-known forcing-out phenomena may occur: substance forces out process, money forces out process, power forces out process. Each of them poses a risk to the process manager's position, which creates a paradoxical situation: the more resources a process manager has available, the greater the risk they will harm his position.

> Rosabeth Moss Kanter lists the skills of the change manager ('change agent'), four of which we quote here:
> 1. The ability to perform effectively, without the power, sanction and support of the management hierarchy;
> 2. The ability to develop high-trust relationships (. . .) ;
> 3. Respect for the process of change, as well as the content;
> 4. The ability to work across business functions and units, to be 'multifaced and ambidextrous'.[53]

Remarkably, formal power is seen as a disadvantage, and substantive expertise and process expertise are equally important.

In such a non-cooperative culture, it may be asked how the process manager can protect his position, and a second strategy presents itself. In essence, this strategy implies that a process manager should adopt a reserved attitude instead.

- He confines his role to a function (managing the process) and so does not assume any other roles.
- This reduces his power to solve problems himself. Where possible, problems and conflicts are left for the parties to solve jointly. After all, it is they that have to support the eventual decision; they can only do so whole-heartedly if they themselves are the owners of the solutions for their conflicts.

The process manager therefore has a limited position of power. He has, however, a number of strategies available that can reinforce his position.

Framing: conflicts between parties

A *project* manager is solution-oriented and may settle conflicts in his team himself. This is a risky course of action for the process manager, since every settlement may create conflicts, causing the *process* manager to become a party in the conflict. Consequently, it is important that the process manager should manage to 'frame' problems in such a way that disputes, if any, are between the parties and so have to be solved by the parties.

Building up credit with the parties
Building up a certain amount of credit with each of the participants is essential: confidence in the process manager's performance and in his person. Credit may be built up by protecting parties at particular moments in the process or by offering them extra room to promote their interests. Credit needs to be built up first and foremost with parties that are able to exit the process without suffering too much harm. Usually, they will require greater protection of their position by the process manager.

Entering into moderate conflicts with parties that have no exit options
Conflicts between the process manager and the parties usually do not enhance the process manager's authority, but cannot always be avoided. Much depends on which party the conflict is started with. Parties with few exit options are candidates. They cannot exit the process after a conflict and have an interest in the process manager keeping the parties with many exit options in the process. This is why they will have, and need to have, a higher pain limit for conflicts with the process manager than the parties with exit options.

Bypasses with parties
A process manager may develop bypasses with particular parties: relations in addition to those between the process manager and the process party. Such a bypass allows communication with a party along a second line. Bypasses make it possible to enter into a conflict with a party. This is because there are more relations between the process manager and the party (the 'bypasses'), permitting a more cooperative relation with the party in question apart from the conflict. In many cases, a process manager has such a bypass with the principal's representatives. There is a line from process manager to party in the process, and a line from process manager to principal. In some cases, a conflict with the representative of the principal in the process may be mitigated by the commissionee-principal line. The extra relationship may make the principal slightly more sensitive to the dilemmas facing the process manager or explain the process manager's attitude.

Controlling external support
If conflicts occur in a process, the process manager may ultimately seek the support of parties outside the process. They may draw the attention of the parties involved and of the process manager to the importance of the process, which may be an incentive for these parties to cooperate. In negotiations between industry and societal organizations, a government may conceivably express its support for the process and the process manager may indicate that unilateral government measures will be taken if the process fails.

5.4.2 *The progress of the process has an independent value ...*

A process manager who behaves in this way keeps the process of consultation and negotiation going. From the perspective of classic management styles – command and control, substance and project management – this is hardly significant. From such a perspective, the notions from section 5.3 – transforming intended substantive decisions into procedural decisions – will hardly be appreciated. These management styles focus on taking efficacious, substantive decisions. It should be observed once more that such an attitude is unlikely to succeed in a network: the stakeholders may block such a decision; the firmer the decision, the stronger the opposition on the part of the stakeholders. Moreover, many of the issues are unstructured problems, for which there is not one, unequivocal, right solution. Consequently, stakeholders blocking a decision always have substantive arguments for their behaviour. Where there is such dependence on stakeholders, it is crucial for a manager to keep the process of consultation and negotiation going.

... because processes produce unfreezing

In the first place, parties may lose their firm convictions; unfreezing takes place. Usually, such unfreezing can only occur as a result of repeated interaction with other parties. Such latent learning effects tend to be underappreciated because they fail to produce sufficient visible results. They are also an important breeding ground for later decision making. Negotiations are possible only when parties are prepared to lay their own views on the table.

... because relations develop and profit potential emerges

Second, relations develop between the parties negotiating in a process. These relations may be very attractive to the parties because they may be used to solve other problems than those on the agenda. In the 'slipstream' of the process, problems hardly related to the process agenda or not related to them at all tend to get solved. In addition, profit potential may, of course, also evolve for parties with regard to the items on the agenda. If so, it will become attractive for these parties to take decisions, allowing them to cash in on their gains.

... because the cost of an exit option will go up

Third, the cost of an exit option keeps rising during the course of the process. A party taking a non-cooperative stance during the process runs risks: it jeopardizes the profit for all parties and impairs its relations with these parties. This may cause the cost of non-cooperative behaviour to skyrocket. As a process progresses, the incentives for cooperative behaviour will therefore increase, and it will become increasingly difficult for parties to back out of the result of a process. The mild pressure exercised by the process thus provides more control than the hard way of command and control.

The pressure to arrive at final decision making will increase during the course of the process. Processes tend to have their own dynamic: once they have been started and are in motion, it is difficult to terminate them halfway.

> Here, we refer to the descriptions of the long-drawn-out negotiations between the ANC and the South African government about abolishing apartheid. Although these negotiations are held in secret and hardly visible for anyone for a long time, the process acquires its own dynamic. The white government believed that it could always exit the process, but at one stage finds that this is no longer possible, although it has to sacrifice a lot more in the negotiations than it wants.[54]

... because there is less potential for starting a conflict

The above advantages will not always manifest themselves. The conflict of interests between the parties may be so severe that no prospects of gain emerge even during the process. Keeping a process going may also have an important fourth advantage.

Interaction between the parties will reduce the risk of an open conflict. The parties may have an interest here: they know that the conflicting interests cannot be bridged, but they also know that the cost of a possible conflict may be high. Keeping an interaction process going may then take on an independent meaning. This mechanism occurs in the negotiations in the Middle East: keeping a process going, even though there are no results, reduces the risk of an open conflict.

... because the environment's expectations form incentives for cooperative behaviour

Fifth, the progress of the process implies that the environment has increasingly high expectations of the outcome of the process. This may be an incentive for the parties to behave cooperatively.

An example: societal organizations and industry negotiate about the question what is the least polluting package. Both parties may at some stage have an interest in delaying this dialogue: industry because it fears that expensive packaging systems are the least polluting; societal organizations because it may be found that they have been campaigning for years for a packaging system that is more polluting than they thought.

The process they started may create high expectations among the public, however. An often-heard complaint is that consumers are confused about what package is the least polluting. This may lead to pressure on the process by particular organizations: the participants will at any rate have to announce some results. The longer the process, the higher the expectations, and the more difficult it will be to use an exit option.

5.4.3 Process agreements: instrumental use or intrinsic value?

This section examines whether the process is a means to realize particular substantive aims or whether compliance with process agreements is an aim in itself. The answer is that both stands may be correct.

Chapter 2 already set out that a purely instrumental approach of process management is risky. In a context of conflicting interests, different actors tend to give different answers to the question whether process agreements will at any time serve a higher purpose. This may cause a party not to comply with particular process agreements, invoking a higher purpose. Even if this party acts in good faith, from the conviction that all parties will agree with it, there is still some chance that other actors will not be convinced of this. They will regard this party's behaviour as a justification for behaving in the same way on a following occasion if they feel that the process agreements no longer serve any higher purpose. As a result, the process as a whole may lose its disciplining value.

Instrumental	**Intrinsic**
1. Realizing change requires support.	1. Realizing change requires support.
2. This is a fact I have to accept. I would have preferred otherwise, because we have a good and well-considered substantive concept, but I must be realistic. Naturally, there is resistance; organizations are not used to change and the psychology of change requires the change to be implemented neither too soon nor without consultation.	2. This is a fact I have to accept. I can, however, also turn it into it something positive. The organizations participating in the process have information, knowledge and insight I do not have. They might improve the substantive concept I devised beforehand, enhancing not only support for it, but also its quality. It also takes some of the weight off my shoulders: I do not have to devise and control everything myself.
3. Some participants came up with new ideas during the process. Interesting, but they had been devised before. Frankly speaking, in the process I am not so much consulting about the change, but explaining again and again how the change should be worked out. Some of the new ideas are in conflict	3. Some participants came up with new ideas during the process. They set me, or rather, us, thinking. Some of the new insights differ markedly from the substantive concept on the table at the start of the process and, frankly, they conflict with the basic principles of this concept. Consultation implies that the

with the basic principles and so can no longer be the subject of discussion.	concept may change and the quality has risen.
4. The transparency and depoliticizing effect of the process approach are important. They lower resistance, allowing the participants to get accustomed to the new concept. It is a smart way of breaking through resistance.	4. The transparency and depoliticizing effect of the process approach are important. They lower resistance and make it possible to build the confidence among participants that their insights and ideas are taken seriously, as long as everyone is prepared to have their own insights and ideas criticized by others.
5. In my view, a process approach is a good instrument if it leads to the introduction of the substantive concept, perhaps less quickly than desired.	5. In my view, a process approach is a good instrument if it leads to the introduction of the substantive concept, perhaps different from my original aims, but better and with wider support.

Table 5.3 An instrumental and an intrinsic approach of process management

The above may be formulated otherwise: compliance with process agreements may be an aim in itself. It improves cooperation, it prevents opportunism and it orders intercourse between parties. It thus prevents the decision making becoming so capricious and unpredictable that a consolidation of decisions is hardly possible any longer. It means that the above positive effects of keeping a process going can be realized. So a process design is more than just a means to realize higher aims. It may be an aim in itself: the stability and predictability of relations.

But there are serious and imminent risks when compliance with process agreements is seen as an aim in itself. An important risk of a process approach as aim is that a process or a number of process agreements will coagulate. As a result, they are not longer monitored critically and no checks take place to see whether the process agreements still serve the purpose for which they were created. Such an attitude may result in strong faith in the process agreements: they are complied with, and compliance is seen as a guarantee that the advantages of the process approach will be realized. In such a situation, process forces out substance.

The building of the Challenger space shuttle may again serve as an example. As we observed earlier, attempts were made within NASA to reduce substantive uncertainty by means of a process approach. Research shows that this process approach is effective, but that the organization tends to lapse into formalism and excessive confidence in process rules and procedures after some years: anyone strictly

> observing process rules and procedures has succeeded in principle. Formal considerations may give rise to questions like: Were the deadlines met? Were the guidelines for the review procedures observed? Does a proposal meet the procedural requirements for numerical arguments? etc. Vaughan, the author of the study referred to, shows that such formalism slipped into the NASA organization over the years. Starting as means, formal procedures tend to become aims: formal considerations force out technically substantive considerations. As a matter of fact, once agreement has been reached in the NASA organization in accordance with the process rules, the result of the process would seem hardly contestable. At NASA, the conclusion of the review procedures was that the O-rings do not always function at low temperatures, but that this is no problem if spare O-rings, *back-ups,* are fitted. Such an incorrect view will become a given within an organization: what, in principle, is merely *negotiated knowledge* becomes objective knowledge.[56]

It may be concluded from the above that process agreements acquire an intrinsic value if the process manager values lasting relations. Some degree of institutionalization of process rules is desirable then. Conceivably, institutionalization poses a threat in some situations. Process may then force out substance. In such a situation, the instrumental function of the process rules is paramount.

Notes

[48] 'Enlarge pie first, cut later' is the mechanism. See for example: L. Susskind, Using Assisted Negotiation to Settle Land Use Disputes: A Guidebook for Public Officials, s.l., 1999; also see: J.E. Innes and D. Booher, "Metropolitan development as a Complex System: A New Approach to Sustainability", in: Economic Development Quarterly (1999), No. 2, pp. 141-156.
[49] De Bruijn, Ten Heuvelhof and In 't Veld (1998).
[50] H.M. Weening, The Flywheel Long Gone? An evaluation of the course of and the approach to the process surrounding the Smart City, Delft, 2001 (in Dutch).
[51] R. Moss Kanter, When Giants Learn to Dance: Mastering the Challenge of Strategy, Management and Careers in the 1990's, London, 1989; Buchanan and Huczyncky (1997).
[52] H.L. Klaassen, Besluitvorming en afhankelijkheid: Over de rol van procesarchitect bij overheidsprojecten, Rotterdam, 1995, pp. 184-185.
[53] Moss Kanter (1989).
[54] Sparks (1995).
[55] Vaughan (1996).

6. PROCESS MANAGEMENT AND THE PROTECTION OF THE PARTIES' CORE VALUES

6.1 INTRODUCTION

Chapter 5 discussed the open nature of decision making. Open decision making has major advantages, but may also be highly menacing to the parties involved. They have particular interests and are not always sure whether they serve their interests by participating in an open decision-making process. They might get captured, or open decision making might produce a result that does not satisfy them.

The second core element of process management is therefore that a secure environment should be created for the parties: they need to be certain that their core values will not be prejudiced. A core value is a value that is vital to a party's existence. Harming this value means harming the party's essence and might rule out its proper functioning. See section 6.2 for a number of relevant examples.

We will work out the idea of protecting core values by answering four questions:

How can the parties' core values be protected? (section 6.2)

What is the nature of the parties' commitment to the result of the process? Should they commit themselves to the result of the process in advance, at the start of the process? Is the result binding on the parties once the process has been completed and the parties have established a result? (section 6.3)

Should the parties commitment themselves to the interim results in the course of the process? (section 6.4)

How should we deal with the exit rules of the process? How can we prevent the parties exiting prematurely, which would harm the process' progress? (section 6.5)

We would like to point out here that most of the answers to these questions are counterintuitive. It would be obvious from a project-management attitude to demand that the parties should commit themselves to the result of the process, that they do so as early as possible and that they should also pledge their commitment to interim results during the process. The parties should also be given as little room as possible to use the exit option. If not, there is hardly any guarantee that a process will be successful and produce actual results that find sufficient support among the parties involved. A process may then be a very laborious and inefficient style of management.

Clearly, such a project-style attitude is hardly effective in the process. The parties might perceive such a process design as very menacing. They will not easily be prepared to participate in a process, and will see it as a trap: it drives them into a particular direction from which they cannot escape. It is far more elegant to allow the parties room and guarantee that their core values will be protected under all circumstances. Room leads to relaxed participation by the parties to a process, as a result of which the process is given a fair chance of being effective.

6.2 PROTECTING CORE VALUES

As a rule, processes that insufficiently protect the core values of the parties involved are unlikely to succeed. We shall give a number of important examples.

- *A politician has political responsibility as his core value:*[57] he must be able at all times to answer for his actions before an elected body (parliament, the city council). A politician who participates in a process might get stuck between the participants in the process on the one hand and parliament on the other hand, since the result of the process may be a decision that is unlikely to receive the support of the majority of parliament. For that reason, participating in a process holds little attraction for a manager; it is much more attractive to him to keep a free hand. This will be different only if the process is designed so as to protect the political manager's core value. This design is set out below.

- In many cases, *companies* are invited to participate in a process, as is clear from the above-mentioned examples of the decision making about the environmental impact of packages and about constructing or not constructing a second Maasvlakte. An important value for companies is the *confidentiality of corporate information.* As a rule, they will hardly feel inclined to submit their figures about the development of particular markets or about the cost structure of their products to third parties. This is a core value for companies, which has to be protected in a process. Companies must be able to rely on it that , for example, they will not be forced by other parties to submit the cost analysis of a number of packages in discussions about the environmental impact of packages.

- *Societal organizations* also have core values. The central mission of many societal organizations is *to inform and to activate* public opinion. Suppose a societal organization is invited to participate in a process on the condition that during that process it will not make public any information or views about the subject at issue in this process. Clearly, such a condition is unacceptable to many societal organizations. One of their core values is to inform the public, and a process must never harm such a core value.

The fact that the parties can be sure that the process will not harm particular core values may be an incentive for cooperative behaviour.
How can these core values be protected? The answer is simple: by establishing rules of the game for it and by also using the process to create respect among the parties for the core values of the other parties.
The risk of this is that the parties will too easily invoke their core values to protect their own position in the process. After all, the above core values can also be used opportunistically; they then form an occasional argument to distance themselves from the result of the process. How to deal with this risk?
Processes nearly always result from negotiations between the parties (see chapters 3 and 4). A party will put forward its core values in these negotiations. The other parties will learn that this party will adopt a cooperative attitude only if its core value is protected in processes. The party with the core value can learn at the same time that other parties fear that this core value will be invoked too lightly. The parties will start negotiations. These tend to lead to an arrangement protecting the core value of the party in question but also holding out incentives to prevent them being invoked lightly.

Every four years, the Ministry of Education and the public transport companies (Dutch Rail and urban and regional carriers) negotiate about a travel scheme for students, allowing them free public transport for one academic year. A key feature of these negotiations is asymmetry in information: The Ministry of Education lacks information about the factual and expected use of the transport scheme (how many kilometres are travelled where and when) and about the costs the public transport companies allocate to transport the scheme. The Ministry of Education will need some insight into these details. The public transport companies may regard these details as confidential corporate information, however. This makes it a core value, deserving protection.
If the parties then negotiate about a process for dealing with this core value, the outcome may be the following:

- The parties will initiate a joint study into particular facts and figures. This study is managed and supervised by the parties jointly to prevent the results being influenced too much by the assumptions and views of one of the parties.
- As regards facts and figures that cannot be obtained from research (for example because they can only be derived from the corporate information of the companies), the parties establish the process that facts and figures that one of the parties has available will be disclosed to the other party where possible. However, none of the parties is obliged to do so.
- If the public transport companies regard particular information as confidential, they will say why. The Ministry of Education will form an opinion about it. If

the parties disagree about it, each has the right to lay down its own view in the public document containing the final result of the negotiation.

- If the public transport companies regard particular information as confidential, the Ministry of Education reserves the right to gather the relevant information by other means.

This creates a process that protects the core value of the companies but that also contains incentives to prevent invoking core values too easily.
If the public transport companies invoke confidentiality, they have to substantiate why they do so (an argumentation that the Ministry of Education may challenge). Keeping open the option of announcing publicly that particular data has not been submitted creates a mild incentive for the public transport companies to be rather generous in supplying information. After all, such a passage would be undesirable if both parties want good results. (It might be a reason for parliament, for example, to be critical of the result of the negotiations).
The agreement that the Ministry of Education is allowed to gather information itself creates a similar, mild incentive. In *bona fide* negotiations, the public transport companies might wish to prevent this. An alternative would be to carry out a joint study or to disclose their corporate information to the Ministry of Education.
Incidentally, chapter 4 explained that a process design is nearly always the result of negotiations between the parties involved. The importance of this is apparent again here. The above process cannot be invented beforehand, but can only result from negotiations.

As will appear later, too, this is one of the characteristics of process management:
there is a core value that is given protection, but the process is such that there are incentives to make limited use of it during the process.

6.3 COMMITMENT TO THE PROCESS AND THE RESULT

What does the protection of the parties' central interests imply for the commitments the parties make to abide by the process and accept its results?

6.3.1 Commitment to the process, not to the results

A process ends with a result. This may be a package deal, for example: a series of decisions, bringing each of the parties involved sufficient gain. Implementing these decisions will then demand cooperation between the parties involved; each party will have to meet its own obligations, thus guaranteeing the gain for the other party.

The first question is whether the parties can be committed to the final result at the *start* of a process. The answer to this question is simple. Processes have an unpredictable dynamic, making it impossible to forecast the final result of the process. That is why the parties cannot expected to commit themselves to the result of the process beforehand. At the start of a process, the parties should always be offered room to distance themselves from the result, if necessary.
If there is no such room, the process will be very laborious. This is because the parties will, first and foremost, want to prevent the process from taking a direction unwelcome to them and will want to influence any interim result in a direction as favourable for them as possible.
The parties' commitment will therefore–of necessity–limit itself to a commitment to the rules of the game: they are willing to join the process if it abides by the processes agreed beforehand.

The second question concerns the nature of the commitment the parties make at the *end* of the process. There is a result then (a series of decisions, for example), which the parties must now implement. Here, too, the mechanism applies that a direct coupling between the result and its implementation poses a major threat to the process. If the parties know during the process that there is a direct link between the results and its implementation, there is a strong incentive for them to enter into close combat and be uncooperative. They will wish to influence the results so strongly that they will run no risk whatsoever that their interests might still be prejudiced in the implementation phase. This will result in very high decision-making costs, and the process might even produce no result at all.
The parties need room in a process, which a direct coupling between the results and their implementation does not offer.

6.3.2 *Offering room to the parties*

The opposite idea is that of indirect coupling.[58] This means that there is always some room between the results of the process and the ensuing implementation.
Such indirect coupling may be an incentive for the parties to show cooperative behaviour in the process. It also reduces the decision-makings costs because the parties are not fully dependent on the results of the process for the realization of their interests. They still have some room–limited though it may be–in the implementation phase. We will set out the advantages of this indirect coupling with the help of an example.

> Suppose two organizations are planning to merge. They decide that this merger will be effected as follows:
>
> 1. A study will be conducted into the opportunities and threats of a merger for the two organizations.

2. Decisions about a merger will then be taken, based on this study: will they merge and, if so, how?
3. These decisions will then be implemented.

The process described under (1) and (2) will be supervised by a process manager. It will involve the main stakeholders of the two organizations.

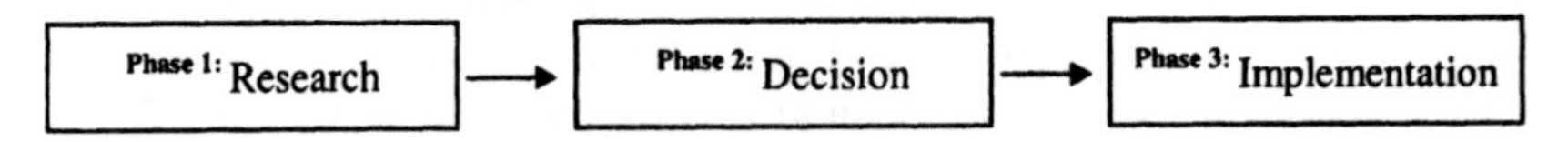

Figure 6.1 Research-decision-implementation

Suppose direct coupling is chosen: the decisions in phase (2) will be implemented exactly in phase (3); the research in phase (1) will strongly influence the decision making in phase (2). Although this might seem attractive, the consequence may be twofold.
First, it may heavily mortgage phase 1. If – as a result of direct coupling – the results of phase 1 prescribe the implementation in phase 3 as almost compulsory, the parties will feel inclined to bring up for discussion all sorts of implementation problems in the research of phase 1. There is every chance that phase 1 will be highly laborious, because the participants may come to feel that the party that loses the game in phase 1 has no further chances. Direct coupling may thus lead to stagnation.
Second, direct coupling naturally does not take dynamic into account. During the process, unforeseen and new developments may occur, the parties may learn, as a result of which – in above example – the research of phase 1 will partly have lost its meaning in phase 3.

... and shape the process in such a way that this room is used as little as possible

Naturally, the risk of this approach is that some parties will distance themselves from the results at the end of the process in which others have invested so much. The idea is, however, that a good process is a strong incentive for the parties not to use this room. During a process, a party will learn to put its own views in perspective (unfreezing) and develop relations and possibilities of making gains. Distancing itself from the result at the end of a process would be putting these advantages at risk. Although the process is then concluded without this party being tied to the result, the parties will, of course, meet again. If the parties have cooperated intensively and trustfully in a process, this behaviour will be a heavy burden on the relations.
Here, too, the dynamic occurs that is typical of process management: offer room at the start of the process and invest in a proper and fair process to minimize the risk of the parties using the room.

> The following situation occurred during the negotiations about the coalition agreement of the first Socialist-Liberal government in the Netherlands. The specialists of the Socialist and Liberal parliamentary parties had struck a deal about a coalition agreement. At the end of the negotiations, Liberal negotiator Frits Bolkestein blocked this agreement. He defended his action by pointing to a rule which–in brief–says: 'Nothing has been decided until everything has been decided'. This rule applied in the trade union world, he said. When asked, the then chairman of the Federation of Netherlands Trade Unions, Mr Stekelenburg, confirmed the existence of this rule, but commented on Mr Bolkestein's action as follows: "Everything remains negotiable, but of course you cannot introduce new elements at the end. Get away".[59]
>
> From a process-analytic perspective, the situation is as follows. The 'Nothing has been decided until everything has been decided' rule offers the participants room. The participants in the negotiation know the rule and it offers them the room to make headway on parts of subjects, knowing that they can review provisional arrangements. Although there is this room on the one hand, the possibilities for using this room at the end of the process will be limited.
>
> In an interview, former prime minister Ruud Lubbers analyses the negotiating process between the heads of government of the countries of the European Union during the 1994 summit in Corfu, where one of the items on the agenda was the presidency of the EU. Mr Lubbers confirms the rule that the choice of a new chairman demands consensus: all member states have to accept the candidate. But he continues: "Of course, a small country tends to join the consensus".
>
> This is another rule that inspires confidence in the participants in the process. They can safely commit themselves to the process, because consensus is necessary in the end and apparently the participants can be forced in a direction they do not like. But that room–it might seem–cannot always be claimed when decisions have to be taken eventually. Small countries in particular have limited possibilities on this point.[60]

It might be added that the parties can make complementary rules of the game about the conditions on which they are allowed to use the room offered to them. A well-known construction is that the parties are allowed to use room if they can produce arguments for doing so. A party may then distance itself from the results of the process, but must substantiate this before the other parties. This reinforces the incentive not to use the room offered lightly: a party that backs out of the results has to account for it, which makes it highly explicit for the other parties why the relations developed with them apparently do not suffice for such a party to commit itself to the results. If a party has gained too little by the process, the obligation to produce arguments makes it possible to justify the refusal to make a commitment.

We will now work out the above in a short discussion of the core value of political managers.

6.3.3 Process management and political responsibility

A minister with political responsibility who participates in a process either directly or indirectly – through an official representative – finds himself in an area of tension.

- On the one hand there is loyalty to the parties in the process; he has passed through a process with these parties and achieved a particular result.
- On the other hand, there is the obligation to give account to parliament. The members of such a body have an obligation to form an opinion – without mandate or instructions and consultations – about proposals made by the minister. They are also a democratically parliament and thus superior to the partners in the process.

This tension is the reason why ministers so often withdraw from a process.[61] How should they deal with this tension? The obligation to give political account may be seen as one of the minister's core values. These core values will have to be reckoned with in the processes between the parties. This may be worked out by the following process:

- the minister participates in the process, either directly or indirectly;
- the other parties give the minister the room to distance himself from a particular result if the parties reach it;
- if the minister uses this option, he will have to produce arguments to the other parties.

The idea is that such a construction meets the minister's obligation to give account to parliament. It safeguards the primacy of politicians. The higher the quality of the process, the weaker the incentive for the minister to use this room. Also, the less room will parliament see for itself to use its right and set aside the results of the process.

He will have to give account to the partners in the process if he uses this room too lightly. The minister may be involved in high costs If they feel that there has been a high-quality process and that he backs out of the results lightly. It will not make his future relations with the partners in the process any easier.

In the event that all partners, including the minister, are pleased with the process and its results but do not receive the support of parliament, the minister is able to give an explanation. He can make clear to his partners in the process that there is force majeure, because the eventual decision lies with parliament.

In this arrangement, parliament maintains the classic position of being the minister's supervisor. Such an arrangement can, of course, give rise to its own form of opportunism:

- *in the relation between the minister and parliament*

For example, the minister may submit a particular measure to parliament for a review, explaining that the measure is the result of a long negotiating process and also forms part of a package of decisions. Rejection of the relevant measure by parliament will frustrate the negotiating partners, who will also shirk their obligations arising from the package. In such a situation, there may hardly be any free choice for parliament.

- *in the relation between the minister and the other parties in the process*

The minister may also make an opportunistic use of the primacy of politicians during the process or at the end of it. This opportunism tends to manifest itself in criticism of the process: he does not take it seriously, calls it unfair, says that it was badly managed, that it has already been overtaken by new developments, that the wrong parties had been selected or that not all alternatives possible have been discussed. The absence of a majority in parliament in favour of the results of the process justifies the minister to the partners in the process.

An attractive arrangement to deal with this tension is the following. In the first instance, the minister consults with the body about the process design before the process. At that stage, the minister is negotiating about the design and parliament is given the room to give its opinion about the envisaged process. The minister, in his turn, can introduce this into the negotiating process about the process design.
During the process, the minister regularly reports to the body. This report not only discusses the substantive progress, but also the process. At the end of the process, the minister gives account to parliament about the process.
This allows parliament to take a considered decision about the measure proposed: how does the past process relate to the process about which the minister informed the body at the start? How harmful is a rejection of the measure by the body to the partners in the process, and thus to the minister's position? The parties in the process can check whether a true picture is given of the process, because account is given in public. Can the minister really distance himself from the results, invoking limited quality and fairness of the process? Is a sufficiently clear statement made saying that rejecting a measure will harm the partners?

6.3.4 The position of the initiator

We pay attention here to the position of the initiator of a process.

A party initiating a process usually has a concrete reason for it. For example, the party involved wants to realize an infrastructural project and knows it is dependent for it on other parties. Such a party then faces a major dilemma.

- On the one hand, this party has to realize its project and thus invites other parties to participate in a process.
- On the other hand, these parties will only participate in this process if the initiator declares that the project is negotiable. Failing to do so would create the impression that the process only serves to realize the project shrewdly.

Take the example of the Second Maasvlakte area. Rotterdam would like to realize a 2,200-hectare Second Maasvlakte, but after some time finds that there is considerable resistance to the construction of the Maasvlakte. Rotterdam will then have to adopt some form of process management: the city can invite other parties to join a consulting and negotiating process. The Second Maasvlakte can form part of these consultations. The opponents of the Maasvlakte, who even have blocking power, will probably demand the idea of a Second Maasvlakte to be open to discussion. If not, the idea of the process hardly appeals to them; their impression will be that the process is a shrewd way to commit them to a second Maasvlakte. Parties with managerial experience also know that the power of a process is very strong: relations and possibilities for gain will evolve, making it very difficult for the parties to back out of the result of such a process (including the Second Maasvlakte). Rotterdam's main rationale for a process is realizing the Second Maasvlakte; Rotterdam would be unlikely to start a process if it had no wish to build it.

What would be an arrangement to solve this dilemma? This arrangement, too, is based on the idea of indirect coupling between the result of the process and its implementation. This arrangement will look as follows:

- The initiator declares that the project it envisages is negotiable.
- The other parties recognize that the project is a sine qua non for the initiator to join the process. They offer the initiator the possibility to distance itself from the results at the end of the process, if and in so far as the initiator produces arguments for it to the other parties.

This arrangement, too, offers room to the initiator: it can distance itself from the result of the process if it fails to bring it the envisaged project. In return for this, it declares the project negotiable at the start of the process. Two scenarios are possible now.

- At the end of the process, the initiator finds that the result does not satisfy it. It may then distance itself from the results. In doing so, it will make the trade-off that – on the one hand – this means that it is jeopardizing the positive results of the process (the package), but that – on the other hand – this gives it a free hand to try to realize its project in a different way.

- At the end of the process, the initiator finds that the result does satisfy it. This may mean that it has realized the envisaged project or has realized it to a sufficient degree. Instead, it may not have realized his project, but it is nevertheless pleased with the results of the process. This is because all sorts of possibilities for gain have evolved during the process, and some of these possibilities were unforeseeable at the start of the process. In total, this gain may be satisfactory to the initiator. It may also have learned during the process that, for example, there are worthwhile alternatives to the project it would like or that resistance to the project is so strong that it stands no reasonable chance of realizing it.

If Rotterdam declares that the Second Maasvlakte is negotiable and then passes through a process with a number of stakeholders, Rotterdam might be highly pleased with the results at the end of this process, even though a 2,200-hectare Second Maasvlakte will not be realized. Rotterdam may have learned that the resistance to such a Maasvlakte is so strong that it cannot reasonably be realized. The parties may also have negotiated in the process and new, unexpected possibilities for gain may have evolved for Rotterdam. Suppose one of the stakeholders (Vlissingen, a port near the Dutch-Belgian border) proposes that the Rotterdam Port Authority should from now on cooperate closely with the Vlissingen Port Authority. Also suppose that there is a new technological development: there is a new type of ship, with a much greater container capacity than its predecessors, and it cannot reach the port of Antwerp because of its draught. So long as the international treaty obligation that the Westerschelde River should be deepened has not been fulfilled, these ships would have to divert to Vlissingen. A rapid and strong development of Vlissingen would then be desirable. Although the actual decision making took a different course, one can imagine that this package is highly attractive to Rotterdam. A strong position for the Port of Rotterdam Authority in Vlissingen can then be exchanged for a smaller Maasvlakte, while Vlissingen can create a distinct profile for itself as a strong competitor of Antwerp thanks to the new technological developments.

6.3.5 Incompatibility and opportunistic use of core values

Finally, we will briefly consider two questions about the notion that parties' core values have to be respected:

- How to deal with incompatible core values? What should be done when the protection of the core values of party X harms the core values of party Y?
- What should be done when the parties adopt strategic behaviour? They define their points of view as core values, thus declaring them non-negotiable.

Suppose two countries enter into peace talks fifteen years after ending a war. During this war, country X has conquered a valley and continues to occupy it. Country X

argues that holding on to this valley is one of its core values: national security would be prejudiced if the valley were handed over to country Y. Country Y, however, argues that the return of the valley is a condition for negotiations. The integrity of its territory is one of its core values that cannot be prejudiced.
Now suppose that a superpower acts as a facilitator in these negotiations. It will have two problems:

- The core values of the one party prejudice the core values of the other party.
- The facilitator may feel that these are not the parties' core values but views that should be negotiable. As regards the latter: the example shows that the facilitator is unlikely to succeed if it tries to convince the parties that they are not core values. Core values tend to be core values when a party regards them as core values. Respecting them may create more goodwill than attempts to make a party abandon them.

The arrangement that may be designed for situations of this kind will always have to seek to give room to the parties. They must be convinced that the process does not threaten what they define as core values.
The parties may be willing to join the process if such room is offered. This room may be offered through an arrangement such as the one set out above: offer the parties the possibility to distance themselves from the results of a process. The process should then be allowed to take its course: unfreeze, form relations and produce possibilities for gain. The process manager may attempt to proceed to decision making once it has become unattractive for the parties to use the possibility of distancing themselves from the results.

Naturally, we have no solution for the stalemates in the above example, but it is clear that – in essence – every solution consists of two parts.

- It is necessary for these parties to join the negotiating process. Only in a process of mutual interaction can they learn how to put their own views into perspective, develop relations with each other and with other parties (superpowers, neighbour states, financial institutions) and can they learn that they have possibilities for gain: more economic growth, security, support from financial institutions, good relations with (regional) superpowers. This does not mean that a process always leads to unfreezing – that would be too naïve – but it does mean that no unfreezing will occur if there is no process at all. Moreover, there is less chance of an armed conflict so long as the process is going on.
- The parties will only join a process if their core values are protected. The promise that this is the case offer them the room to join the process. This, too, may be realized by offering the parties the possibility to distance themselves from the results if their own core values are not sufficiently guaranteed. The process may be, for example, that the parties have the right to distance themselves from the final result if insufficiently honours their views about the

position of the valley. The process should then take its course: if it leads to sufficient unfreezing, relations and gain, the parties are unlikely to adopt a different view about the position of the valley.

The example also shows how conflicting core values can be dealt with: they can be defined in a process-based way if they are substantively incompatible. In the example, the agreement will not be that party X will always be allowed to retain the valley, but the agreement will be that party X may distance itself from the results of the process if it feels that these results take insufficient account of its views about the position of the valley. The same agreement can be made with party Y.

A similar arrangement can be made with parties that define their own views as core values. In budget negotiations, a party may argue that obtaining a budget of 100 is one of its core values and that it will not negotiate if this is not guaranteed. If the other parties feel that this is a view rather than a core value, it would be unwise to agree that A will always be allocated a budget of at least 100.
A process arrangement would be wiser: party X will be allowed to distance itself from the result of the negotiations if – at the end of the process – it feels that it has been allocated an insufficient budget. It would be unwise, however, to offer such arrangements too liberally. The other parties will not make them until they have learned in negotiating about the processes that X is inexorable on this point. In return for this promise, X will have to make some concession to the other parties.

The problem of the conflicting core values thus shows one of the characteristics of process management: offer room to the parties, let the process take its course, limiting the actual use of this room, and make a process for difficult, substantive problems.

6.4 POSTPONING COMMITMENTS DURING THE PROCESS…

A large number of decisions are taken during a process. The question is how strongly the parties have to commit themselves to these interim results. The intuitive answer will often be that such commitments are necessary. The decision making will thus be funnelled (particular options are ruled out, for example), initial results are visible at any rate and the parties are slowly but certainly forced into a particular direction. As we said earlier, this might seem energetic, but may in reality be threatening for the parties in a process. These parties are dependent on each other. As a result, such an approach carries the risk of a stalemate between the parties.
Opposed to this approach is the principle that commitments may be postponed: when a substantive decision is being taken, a party does not have to commit itself

fully to it. Applying this design principle prevents the parties associating the process with a trap, where every decision constitutes a *point of no return*. It offers room to the parties involved and thus fits into a process in which uncertainties and distrust tend to set the tone.
So the position here is the same as in section 6.3: offering room is necessary for the progress and the quality of a process. A process manager who primarily strives to take decisions and solve problems as soon as possible will be unable to cash in on a number of positive effects of this design principle.

6.4.1 ... reduces the decision-making costs

Some of the decision-making processes about several issues will always be unpredictable. It is unclear during the process what the final decisions will be like. If a substantive decision is submitted in such a situation, it is likely to be impossible for the parties to foresee what these interim results will mean for the final decision making and for their own position.
This may lead to a form of decision making in which the parties want to secure their positions as much as possible. The consequence might be that long negotiations have to be conducted to ensure that the decision safeguards the parties' interests as much as possible. In turn, the consequence of this may be that the decision is either complex and detailed or sketchy and vague. Postponing commitments may forestall such laborious decision-making.

A familiar example of postponing commitments is that a party with respect to particular decisions notes that it 'does not feel committed yet', or notes that 'the decision may be reviewed should new information become available'.
These and similar forms of postponing commitment keep the decision-making costs low. The decision making would probably take a very long time if a binding decision had to be taken about these subjects on each occasion because of the uncertainty about the consequences for the final decision making.

6.4.2 ... offers possibilities for dealing with the capriciousness of the decision making

It may be added here that–in complex decision-making processes–a great many decisions are taken that will not eventually influence the final result. What may seem a main principle at the start of a process may become a mere detail after some time. A reverse movement is also possible: in retrospect, a detail may prove to be a crucial decision.
In a capricious and unstructured decision-making process, offering room is a contingent design principle. On the one hand, the design principle prevents a great deal of energy getting lost in settling differences of opinion that prove to be details later. If, on the other hand, irreversible decisions are taken in the opening

phase about alleged details (which will probably provoke less resistance), they may condition the rest of the decision-making process in an undesirable way.

6.4.3 ... offers possibilities for building mutual confidence

Process designs are made to bring the parties together. In such a context, there is almost automatic distrust.
Such distrust tends to lead to a quest among the parties for the interests and hidden motives of the other parties by the time a decision has to be taken. Why does party X take this stance? What is a party's hidden agenda? Rumours and suspicions may then begin to dominate, which may seriously harm the process.
If the parties are uncertain and/or suspicious, but are not forced to take a stand because they can postpone commitments, a certain confidence may grow, both in the process manager and in the other parties.
Thus, postponing commitments is a vital breeding ground for mutual confidence and thus for a successful process and may in the course of time reduce uncertainties and strategic behaviour. A process manager who offers room invests in a cooperative attitude of the parties. Paradoxically, offering room to strategic behaviour may cause it to diminish after some time (see chapter 8 for the problematic nature of the concept of strategic behaviour).

6.4.4 ... furthers learning processes

It is important that much of a process should be a learning process through which the parties pass together (!). During the process, new insights will become available, facts turn out to be different from what they are often thought to be and even normative views can change. Such a learning process can be seriously blocked if binding statements are made at an early stage.
As we pointed out earlier, the parties tend to have no common frame of reference at the start of a process. When these parties are brought together in a process, their belief that only they are in the right will often have to be put into perspective first, before a common frame of reference can be built up. This requires a learning process, which can be realized thanks to the very room created by postponing commitments. A frame of reference comprises a number of dominant ideas among the parties about the course of the process, its substance, its interim and final products, etc.
This often implies that – by postponing commitments – behind a front of 'non-decision-making' there is a reverse of learning processes and the structure of a common frame of reference. Such a frame of reference may be a breeding ground for rapid decision making later in the process.

> If a country has to take an important infrastructural decision – for example about its future airport infrastructure – a dialogue will be organized between the interested

parties. This dialogue leads to a particular outcome: the parties may have reached agreement about particular issues, while the differences about other subjects can be formulated better than before.
In such a dialogue, the tension between economics and ecology tends to be one of the main points of discussion. The process manager's aim may be that the parties not only show 'participative openness' (the points of view are exchanged), but also 'reflective openness': they are willing to bring their own views and values up for discussion.[62] This might break the stalemate between economics and ecology on a number of points.
Such a 'reflective openness' implies, however, that–at times–the parties involved have to abandon deep-rooted views, which is not without risk. After all, the other side might use this for its own purposes ("environmental movement backs more flights"; "industry admits economic advantage unclear").
Suppose it has been agreed beforehand that the result of the dialogue will have a direct influence on the political decision making. In other words, there is a hard coupling between the dialogue and the decision making. This may make the parties scared of bringing their own views up for discussion. The chances are that this will be made improper use of in the political decision making. The result may be that there is no 'reflective openness' in the dialogue, but only 'participative openness'. Indirect coupling, however, may produce 'reflective openness'.

As regards the learning effects, it is also important that a process should move from substantive variety to selection (also see section 3.6). A variety of options should be considered in the opening phase of the process, allowing an authoritative selection to be made in the closing phase. A development from variety to selection cannot be reconciled with too rapid decision making at the start of the process.

A learning process also for the process manager
The process manager plays a prominent role in building up a frame of reference. Postponing commitments may bring confidence, which tends to be a major condition for the process manager to gain an insight into the interests and sensitivities of the various parties in the process. This information may be of great value for the process manager later in the process, for example, when possibilities for gain have to become available to each party.

6.4.5 ... relieves the decision making

In the course of the process, the results have to be formulated, made presentable and communicated. Open decision-making processes by definition generate a great deal of information, the volume of which has to be reduced in the course of the process.

This implies that the aggregation level at which the available information must be processed will rise sharply in the course of the process. This will diminish the relevance of all sorts of subdecisions. Communicating the results externally also requires a certain simplicity and unambiguity, which in the tail of a process is an incentive for the parties to reach decisions.[63] This is another reason not to take substantive decisions too early: given the aggregation level at which the information will eventually have to be processed, substantive decisions may have little impact on the final result.
Table 6. 1 summarizes the above. It also lists the consequences of the opposite attitude, which seeks to take a maximum number of 'binding' decisions in a minimum of time

Design principle: process manager offers room	**Design principle: process manager formulates preconditions and is wedded to consolidation**
Helps to create confidence between the parties	Helps to create distrust between the parties
May diminish strategic behaviour	May intensify strategic behaviour
Improves the progress of the process, which is a value in itself	Increases the risk that the process may be halted, which may greatly hamper the start of a new process
Starting processes by generating variety; room will then improve the quality	Starting processes by generating variety decision making can be conditioned in an undesirable way Decision making implies unnecessary loss of managerial energy
The process manager can learn what the parties' sensitivities are and thus estimate the scope for win-win packages	The process manager loses eyes and ears and can fill his role less and less effectively
Decisions can be taken in an relaxed atmosphere and lessons can be learned	Decisions are taken in a cramped atmosphere hardly conducive to decision making

Table 6. 1 Room and preconditions compared

Of course, a major risk of this design principle that no result of decision making will be visible for a long time. Chapter 7 will discuss the question how to deal with this.

6.5 THE EXIT RULES OF THE PROCESS

An important design principle is that a process should have exit rules. It may be included in the processes, for example, that the parties may consider after some time whether they wish to continue their participation in the process. This may be

an important agreement, which may lower the threshold for particular parties to join the process.
The process manager will also make every attempt to prevent the parties actually exiting. Ideally, the process is so attractive to the parties after some time that exiting is no longer an attractive option for the parties. Here, too, the mechanism is that the parties must be offered room, but that the process must then be so attractive that this room is not used. We will make a few complementary observations.

6.5.1 The participation paradox: an exit option may be attractive

The aim of involving the parties is to improve the quality of and the support for the envisaged decisions. Paradoxically, the opposite effect may be reached. Particular parties participate in the process and thus obtain more and better information than if they did not participate. This information is even used at the end of the process to combat the decision making rather than supporting it. The information obtained can enable better and perhaps more convinced resistance to the decision making than if the party involved had not participated.
A party displaying this behaviour at the end of a process may be blamed for opportunism. This is why it may be attractive for a party to use the exit option during the process and then fight the decision making. There is less risk then that this party will be blamed for opportunism (after all, it uses a right), while it is able to use the obtained information to fight the decision making.

6.5.2 Threatening to exit: double binds for the process manager

If a party threatens to exit, a process manager is faced with a number of risky 'double binds':

- *as regards the party that wants to exit*

The party that wants to exit the process will wish to do so because the process has not got enough to offer it. If such a party wins and can therefore exit, the process will be harmed. If such a party loses, it may believe even more strongly that the process has little to offer, which may also harm the process.

- *as regards the other parties*

The other parties will closely follow the actions taken by the process manager. If the process manager loses and the party exits the process, the process manager will be harmed. If the process manager wins, the first double bind can become effective: the losing party believes even more strongly that the process has little to offer. This will only be different if the process manager can hold out to it a prospect of gain, which, however, may cause the other parties to believe that threatening to exit will pay.

This, too, hardly furthers the progress of the process.

An imminent exit is hard to manage: others have to do this
It is important that the process manager can hardly manage a threatening exit: it is the participant's right. Another mechanism of process management returns here: frame problems in such a way that they are a conflict between the parties and not between the process manager and one of the parties.
There is a major risk that the latter will happen, because an exiting party harms the process manager's key interest: it threatens the progress of the process. In such a conflict, the process manager will always be faced with the above-mentioned double binds, which make any choice made by the process manager highly risky.
However, if the conflict can be framed as a conflict between the parties, the process manager can again play an independent role as a facilitator in the conflict. In so far as the process manager can influence the conflict, he will do so indirectly: through other parties in the process or through the process environment.

Notes

[57] Edelenbos (2000).
[58] C. Perrow, Normal Accidents, Princeton, 1984.
[59] NRC Handelsblad, 30-6-1994.
[60] idem
[61] Edelenbos (2000).
[62] P. Senge, The Fifth Discipline: The Art and Practice of the Learning Organization, London, 1992.
[63] See the description by J. Burger et al., "Science, Policy Stakeholders, and Fish Consumption Authorities: Developing a Fish Fact Sheet for the Savannah River", in: Environmental Management (2001), No. 4, pp. 501-514.

7. THE PROCESS MANAGER AND THE SPEED OF DECISION MAKING

7.1 INTRODUCTION

Process management is marked by openness: the main stakeholders are invited to take part in a process and can participate in fixing the agenda. Openness is not without risk for these stakeholders, however. They can see the process as a trap: once they have joined, they will be forced in a particular direction without being able to exit the process. This is why it is important to protect the parties' core values. At crucial moments, they are offered room for the sake of these core values: for example, they do not have to commit themselves ex ante to the result of the process and are offered the exit option.

This was the argument of chapters 5 and 6. The design principles "openness" and "protection of core values" automatically prompt the question what guarantees sufficient speed in the process. After all, in the worst case there will be many parties presenting many agenda items, which makes the process difficult to control. If these parties also put forward many core values, which they declare to be not negotiable, there is every risk that the process will be very slow or even become deadlocked.

Clearly, networks will hardly have room for classic, project-type mechanisms to raise the speed. In a *project*, fixing a deadline, for example, provides an incentive for speed, but in a *process* this may stimulate resistance. This is because some parties have no interest in decision making before the deadline expires as this harms their interests. Command and control stands hardly any chance, since all parties are mutually dependent.

7.1.1 The dynamic of a process

The main incentives for the speed of a process present themselves when a process is attractive for the parties. A process must be sufficiently attractive at the start (partly as a result of the work of the process architect), but new possibilities for gain should also present themselves during the process. This dynamic develops as follows:

1. Attractive start

Parties are invited to participate in a process and can partly influence the agenda. They will place items on the agenda that are attractive to them. Other parties will do this as well, which results in a `multi-issue' agenda. This multi-issue agenda makes a process attractive, because it permits couplings between the various items.

2. Forming relations

Relations will then evolve between the parties in the process. These relations can be used to take decisions about the items on the agenda. The relations can also be used, however, to discuss completely different subjects, bypassing the process. Parties tend to solve many problems in the slipstream of the process.

Interaction processes also tend to contribute substantively to the unfreezing of parties. Their own views, rigid though they may have been at the start of the process, are put into perspective. This is important as well, because it allows parties to exchange their own subjects for or couple them to other subjects. A process can only result in a loss as long as parties' views are rigid. If a party's views are put into perspective, that party will be convinced more easily that a particular decision or package of decisions will bring it gain.

3. Insight into the possibilities for gain

Possibilities for gain present themselves to the parties as the process progresses. They see possibilities of realizing the wishes they placed on the agenda. They may be unable to realize these wishes in full. But gains may be made even then, because they have learned that there is insufficient support for particular wishes among the other parties.

4. Profits are paid late

To protect their core values, parties are allowed to postpone their commitment to the interim results of the process. As a result, a large number of problems that are difficult to solve may be left at the end of the process.

Postponing commitments also means that a great many possibilities for gain are open to the parties at the end of the process. The reason is that postponing commitments to interim results means that the profit particular parties expect of these interim results has not been paid yet.

This is in compliance with one of the golden rules in the literature about network management: parties' gains should never be paid too early, because important incentives for cooperative behaviour may then get lost.

5. End: profit and loss account

This creates a process that at some stage offers the parties involved relations and gains on the one hand, whereas on the other hand there are a number of intractable problems, about which there is no agreement yet. Each party will draw up a profit and loss account. On the positive side are the relations gained and the profits scored, on the negative side the losses and the intractable problems. These unsolved problems are uninteresting for particular parties (they have no interest in the problem) and a form of loss for other parties (they have an interest in a particular solution of this problem).

6. Profit-loss positive if there is a critical mass: speed.
The speed of the process will increase if this profit and loss account shows a positive balance for a critical mass of parties. They wish to collect their profits and therefore take final decisions. Another important psychological mechanism here is that parties tend to take their profits at an earlier stage.
It is clear from the above, however, that the end of a process is difficult to predict. Ideally, the parties should have some degrees of freedom in fixing the deadline of a process and be able to control it in mutual consultation, allowing a conclusive decision to be taken when there is momentum for it: there are enough positive profit and loss accounts among enough parties to take a final decision.
Ideally, a process should have the dynamic described above. We will now elaborate this basic idea by giving attention to a number of additional mechanisms. The process manager can manage the process in such a way that new incentives for cooperative behaviour continue to be created (section 7. 2). The process manager can use the staffing of the process (section 7. 3). Making proper use of the possibilities in the process environment will speed up the process: environment management (section 7. 4). The way a process is organized may speed it up (section 7. 5). In conclusion, section 7. 6 will discuss the possibilities of providing steering in a process through command and control. Although a command and control style conflicts with the idea of process management, situations may occur during (and thanks to) the process in which command and control may succeed.

7.2 INCENTIVES FOR COOPERATIVE BEHAVIOUR

A process manager has four possibilities of creating incentives for cooperative behaviour. He can do so by:

- drawing up the agenda;
- planning activities;
- having third parties intervene, which may multidimensionalize or reframe conflicts;
- giving parties a repetitive chance of advancing their interest.

The idea is therefore that these incentives will promote the process dynamic described above. We will explain them using the following example.

> Party X and party Y negotiate about the most suitable location for a national airport: Schiphol, the present location in the country's economic heartland, Flyland, an airport in the sea, or a location in one of the country's economically disadvantaged regions . A study is conducted into the effects of building the airport at each of these three locations, focusing on – among other items – the cost, the economic and ecological effects, and the technical possibilities and risks. The parties must submit a

feasible and well-founded proposal for a choice of location. Party X wants to reach conclusions and arrive at decisions quickly and takes the view that the economic effects are paramount. Party Y values a careful study and careful decision making, taking the view that the ecological effects are paramount. How can incentives for cooperative behaviour be created here?

7.2.1 Drawing up the agenda: a balanced distribution of productive power and blocking power

As was observed earlier, the process agenda should be a multi-issue agenda where possible. Only then can parties exchange and couple subjects and will there be strong incentives for cooperative behaviour. It is also very important that this agenda should be drawn up so as to provide an incentive for all parties to use their productive power. Productive power is positive. Parties use their resources to create something. The opposite of it is blocking power: parties block or hamper the decision making, without making a positive contribution to it. If one of the parties only has an incentive to use its blocking power, it will either hardly cooperate or not cooperate at all.

For party Y, there is hardly any such incentive for productive power in the above example. This may be different if the agenda is framed differently. An example of this is the question how, if the airport is moved, the room that becomes available can be used so as to enhance the ecological structure of the area.

7.2.2 Planning of activities

Sequential connection of activities

A second incentive for cooperative behaviour is created by linking the activities to be performed intelligently over a certain period. If party X feels that decisions have to be made quickly and party Y opts for making them carefully, the following structure may be planned: round 1 serves the interest of party X and round 2 that of party Y.

This may create an incentive for party X to take the interests of party Y into account already in round 1 where possible. After all, the more these interests are taken into account, the more chance that the decision making in round 2 will be quick.

The process manager can launch the following proposal, for example: "A quick scan will be made of the economic and ecological effects of the choice of each of the locations. Once the results of this quick scan are known, the parties can form an initial judgement about these effects and formulate additional questions. These questions have to be answered in a new study." This proposal recognizes the interest of quick decision making: a quick scan is started rather than a detailed study. It also

> meets party Y's views, however. If, after the quick scan, party Y should feel that it lacks quality, it may demand an additional study. This may then be an incentive for moderate behaviour by party X. Although a quick scan is made, it should have sufficient quality, or else party Y will receive a great deal of room to demand an additional study.

The agreement thus first meets the interest of party X, but X knows that Y's interest will be served in the next phase. This is an incentive for party X to reckon with party Y's interest. Should party X fail to do so, Y will have a strong incentive to use its blocking power, which will cause a delay.

Parallel connection of activities

In the above example, there is an incentive for moderate behaviour by party X in particular. This is different when sequential connections are made for various simultaneous activities. The structure may be as follows: the activities serving the interests of party X are performed; party X as well as party Y are involved. The activities serving the interests of party Y are performed simultaneously; here, too, both party X and party Y are involved.

> Suppose party X is particularly interested in the economic study into the new Flyland, and that party Y is more interested in the ecological study. The economic and ecological studies may be connected in parallel. The fact that the studies in which both parties are interested run parallel provides incentives for cooperative behaviour. A party that is unreasonably critical of a study that serves the interest of the other party will invite the same attitude in the other party towards the other study. Parties may adopt a moderate attitude as a result of this mechanism.

The mechanism here is that different activities meeting the interests of different parties are connected in parallel. This creates an incentive for cooperative behaviour.

For both plans, the criterion for the way of planning is the extent to which the plan creates incentives for cooperative behaviour. The underlying idea is that a cooperative attitude increases the speed of a process. These mechanisms can never be applied mechanistically, however. Parties may behave strategically, after all. If the planning is sequential, party Y may want to have its cake and eat it in round 1, for example, and subsequently make extra demands in round 2. So there is no mechanistic application; the criterion always applied is whether a particular plan creates an incentive for cooperation in a concrete situation.

7.2.3 Intervention by a third party: multidimensionalizing or reframing

Intervention by a third party may also form an important incentive for cooperative behaviour. The idea then is that such a third party can give a conflict extra dimensions (multidimensionalizing) or can reformulate a conflict (reframing), which may create new room for negotiations.[64]

For this purpose, a process manager may invite parties to formulate a mutual conflict as accurately as possible and then submit it to a third party. This third party will give a judgement about the conflict. Agreements can be made as to what parties should do with the third party's judgement. They may range from *'parties will accept the judgement of the third party in advance'* to *'parties will take the judgement on board in their further consultations'*.

The latter agreement would seem to lack force, but may nevertheless prove meaningful in practice. This is because the judgement need not imply that any of the parties is right. It may also imply that the third party multidimensionalizes: it shows that more dimensions are relevant in the conflict, which may create new room for solutions between the parties.

> Parties have a conflict about whether building an island in the sea at a particular location will be harmful because the current may shift. Party X believes that the present plans can do no damage, party Y feels that the risk of damage is too great. The third party may give a judgement about the variables influencing the current, such as the location, the place of sand dredging to build the island and compensating technical measures. This may allow the process between X and Y to continue. The deadlock has been multidimensionalized, which creates new room for negotiations.

The third party may also reframe a conflict: the conflict is expressed in a different language, which creates new room for negotiations. The first form of reframing is reformulating a conflict in negotiations into a research problem. This implies that a process manager will again invite the parties to formulate their conflict as accurately as possible.

With the help of a third party, this conflict is then translated into a research problem, which is submitted to the independent third party.

> Parties have a conflict about the safety of Flyland. Party X has become convinced that the large number of birds in the sea may seriously threaten the safety of incoming and departing aircraft. According to party Y, this is not so.
>
> They can reframe this conflict into a research problem. This first of all requires accuracy on the part of the parties; they must formulate a number of research questions, such as:
>
> - Where are the large bird populations in the sea and at what time are they there?
> - How regular and predictable is the migration of these birds?
> - How will the construction of an airport affect the behaviour of these birds?
> - What measures can be taken to chase birds away?
> - How do birds react to what measures?
> - How great are these uncertainties in this study?

These questions may then be submitted to researchers, in accordance with a protocol drawn up by the parties. This protocol may imply, for example, that the above questions are answered, but that the researchers also indicate which research findings are 'hard' and which of them contain uncertainties. The result may then help the parties on in their negotiations. The research may show, for example, that bird migration is difficult to predict, but that there are many technical possibilities of keeping birds at arm's length. These solutions cost money, which creates new room in the discussion: the question is not just whether it is 'safe or unsafe', but also how much money the parties wish to invest in risk-reducing measures and the cost-effectiveness of those measures.

The idea is therefore that the study by third parties may facilitate the negotiations. Of course, the fact that parties *can use* this arrangement is just as important as the actual use of it. If a party knows that a particular stance may be the subject of study at some stage, this may be an incentive for this party to adopt a moderate attitude and in its stances recognize the facts as much as possible.
This form of intervention breaks a social fixation by a cognitive intervention. Another form of reframing uses the reverse mechanism. A social intervention breaks a cognitive fixation; a conflict about a research problem may be reframed as a subject of negotiation. Suppose the parties have a conflict about some of the findings of the study into the birds. Reframing implies that they convert this conflict into a subject of negotiation.

For example, the parties disagree about the annual migration of birds. Do they always follow the same route or can there be great differences between the annual routes? This requires a few years of research, which is impossible according to X, but necessary according to Y.
As long as the conflict concerns the study, it may end in a deadlock. Reframing it into a subject of negotiation may imply that X and Y assume that a worst case scenario comes true with respect to the bird migration and that they will then start looking for technical possibilities of keeping the birds at a sufficient distance from the airport. They start consultations about it, which circumvents the research problem.

7.2.4 *Repetitive chance of advancing one's own interests*

As a rule, processes are marked by a multi-issue agenda. Eventually, the parties will have to reach a decision about each issue. It is important that the process design should offer every party a chance plus a number of further chances to advance its interest with respect to this issue. This inspires in a party the confidence that a process design is sound and also prevents it entering the battle at every decision. After all, if a party has only one chance, it will leave no stone unturned to make the most of it when the chance presents itself, which will be a driver for conflicts in the process.

A process agreement may be:

1. Parties have research institute X examine the influence of birds on the safety of the airport in the sea.
2. If this research lacks authority for the parties, they may subject it to a peer review by research institute Y.
3. If, given the comment by peer Y and the reaction to this by research institute X, the study still fails to fully satisfy the parties, they can have additional research done.
4. The research is concluded once there is a decision about the influence of birds on safety supported by research institute A and the parties. If new circumstances present themselves later, this may be a reason for further research.

Conditions may then be attached to these agreements, for example by agreeing that the support of a minority of the parties will suffice for arrangement (2), whereas a majority is necessary for arrangement (3) and a qualified majority is required for arrangement (4).

This process agreement would seem to be time-consuming, but the essence is that it offers sufficient further chances to a party entering the process. If, in addition, a number of further chances are possible *for each issue*, a party has sufficient room to advance its interests. Here, too, it is important to distinguish between offering this repetitive chance and using it. The process will have to take its course (also see chapter 6): if relations between parties and prospects of gain develop during the process, incentives will arise to make a limited use of the room offered.

7.3 THE PROCESS HAS PARTICIPANTS WITH COMMITMENT POWER

The idea behind the need for representatives with commitment power is described in chapter 3: it promotes the external authority of the process and the parties' commitment.

7.3.1 Having representatives with commitment power offers possibilities for gain and incentives for cooperative behaviour

It may be added here that a delegation with commitment power also offers extra possibilities of generating gains and incentives for cooperative behaviour. This is because representatives with commitment power have extensive networks, which

make participation in the process attractive: parts of the extensive networks of the other parties become available to a participant.
Delegations with commitment power also present more possibilities of concluding win-win package deals, since representatives with commitment power have more room to negotiate (they are less hampered by and need less consultation with those represented). This is one of the reasons why it is easier for them to accept a loss than a light-weight delegation, tied to those represented by mandate and consultation.

> From 1990 to 1994, decision making took place about the Per+ project, a large investment project at Shell Nederland Raffinaderij B.V. This decision making is strongly interactive: Shell involves the main stakeholders in the decision making, in which two key figures play a role: Jeroen van der Veer and Hans van der Vlist, member of the Committee of Directors of the Royal Shell Group and administrative officer for the province of South Holland, respectively. "They have, each in their own area, organised and consolidated the required commitment within their organisation."[65]

Another advantage of extensive networks is that more issues can be included in a package. This is because they offer extra possibilities of coupling problems and solutions. A number of possible risks of a representation with commitment power counterbalance these advantages.

- Representation with commitment power implies that participants are difficult to control, precisely because they have these networks of connections. Participants may have other interests beside those of proper speed of the process. The resources they have may then be used against the process: they can use their networks to influence the process through third parties, for example.
- On the one hand, the participants' networks are an important supply channel of information. The most recent information is always at hand. On the other hand, these supply channels can make the process manager dependent. They can be used to disseminate information strategically, hold it back temporarily, blow up details, et cetera.

Consequently, this *paradox of representation* with commitment power means that an important explanation for a successful process (the participation of the main stakeholders at a high level) is also the main threat to the process (the heavier the weight of the representatives, the greater their destructive power).
This paradox requires some form of control by the process manager. This is not easy, because representatives with commitment power are, as a rule, difficult to control. The core of such control is that the representatives will also be dependent on each other in other places and in the future.

In the above example of the Per+ project, Shell is interested in a good investment climate in the area and so in good relations with the government. In its environmental policy, the government will increasingly and more frequently have to rely on the self-rule of companies and is therefore interested in good relations as well.[66]

This may have a moderating influence. The role of the process manager is the same as the one described in the previous section: identify and specify these dependencies, hoping that they will automatically allow the process to take its course.

7.4 MANAGEMENT OF THE ENVIRONMENT

A process has an environment: actors who do not participate in the process, but have an interest in the process or in its outcome. The process manager can use them to gain support for the process, but these actors may also threaten the process. The process manager has to be alert to this and use the environment of the process.

Management of expectations
If the parties outside the process have high expectations of the process, this may be an important incentive for cooperative behaviour by the parties in the process: if the environment has high expectations, it is unattractive to be pictured as the party responsible for the failure to meet these expectations.
As the process proceeds, expectations may increase and it may become more difficult for the parties in the process to get round them. Here, too, the process takes its course: relations between parties outside and inside the process can be used to fasten the process.

See, for example, the negotiating process between industry and societal organizations about the environmental impact of packages. Politicians and administrators as well as scientists eagerly await the outcome of the process. In other words, if the steering committee finishes its job properly, it will have something to offer to its environment. How effective the LCA method was may be made clear to the world of science. New insights emerged in the process, (e.g. about the economic examination of the chain), which are interesting to the scientific community. Naturally, the government is interested in the outcome of the process; apart from the Ministry of Housing, Spatial Planning and the Environment, the EU is interested as well. International industry takes an interest in the process, partly as a result of similar policy developments in other countries.[67]

The threat of a low product-time ratio

An important problem for the process manager is that pressure from the environment can only be generated if this environment regularly sees results of the process. If the parties in a process fail to generate products, the outside world may start to believe that the process is merely sluggish and torpid, which may harm confidence in it. Interactive processes are slow, however. Too slow, may be the environment's reproach.

The objection of the low product-time ratio may be refuted by pointing out that the products that are eventually delivered demonstrate that the decision making as a whole was not slow. Time is a relative concept: time tends to be lost at the start of the process, and this loss is made good at the end of the process.
All the same, an important problem presents itself here. It can only become clear *after* the process that the product-time ratio is high. *During* the process, however, the process manager will have to live with a low product-time ratio, while being unable to guarantee, or able to give limited guarantees only, that he will complete the process successfully, allowing the product-time ratio to rise.

> These problems may occur especially at the start of a process. A process is based on interaction: parties exchange views, negotiate and learn about them. As a result, unfreezing may take place: during the design process, parties become uncertain about their views, information and/or aims. This is a positive development. Only when parties learn how to put their own views into perspective can room be created for negotiation and decision making.
> Unfreezing may be a problem for a process manager, however. Those represented may start to believe that the process approach will be unsuccessful: unfreezing takes time, it seems as if there is no speed, while it may also be a problem for those represented that their representative has doubts about his own views. In short, it looks as if the process lacks effect. Figure 7. 1 presents a schematic representation of the risk.
> The dotted curve shows the course of a process of change as desired by the parties: from a situation of limited consensus to one of sufficient consensus to arrive at decision making. The interrupted curve shows the actual level of consensus during the decision-making process.
> When the curve falls below the X axis, the parties may get the impression that a process approach does not work: there is too much divergence and too little speed. Actually, the process has a kind of incubation period, in which unfreezing takes place. This incubation period is a sine qua non for a proper process, but also makes it vulnerable. The process manager has to make clear to the parties in the opening phase – which is therefore a critical phase – that the process does observe the intentions of a process design.

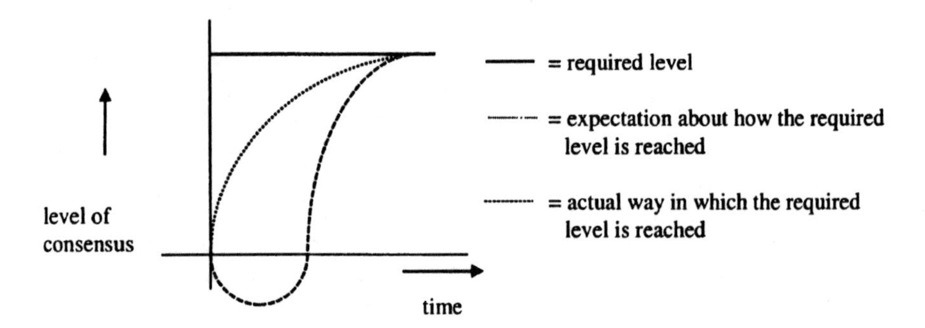

Figure 7.1 The incubation period of a process approach

Impression management means that the process manager ensures that the environment has a positive impression of the process. The most ideal situation for the process manager is that there is a general feeling that a process approach is 'the only way for parties to reach a decision'.
An important aspect of impression management is that the process manager ensures that there are interim products, showing that the process is making headway and is more than the Echternach procession. It may make sense in particular processes to agree with the participants that there will be a number of initial results very soon, not so much because of the results, but because of the impression the environment has of the process. Of course, quick results may also increase the participants' support for the process.
These results tend to be the subjects to which there is relatively little resistance. This is a sensible strategy, particularly in processes that will clearly have a long run.[68] This makes the criterion as to when a process should result in a product mostly process-based: when it is desirable for the sake of the external legitimacy of the process.

Initiatives are being taken in many countries to set up *smart cities*. These are districts with an advanced information infrastructure, offering a multitude of services. Setting up such a district needs the cooperation of many parties (municipalities, residents, providers of infrastructure and services, et cetera), facing a great many risks (administrative, technical and economic ones). Setting up a smart city requires a consulting and negotiating process between these parties. This process should make the parties jointly estimate the risks and also jointly plan the organization of the smart city so as to be sufficiently attractive to all parties or a majority of them.

Regular initial results are very important in the design and management of such a process. This is an important incentive for parties' cooperative behaviour and assures external financiers that the process is productive and thus also a variable increasing the speed of the process. Without such interim results, there is a major risk that a process will get bogged down or the smart city will lose momentum.[69]

A consequence of impression management is *ambition management*. It is important that regular communication with the environment should prevent it from having expectations of the products of the process that are either too low or too high. Parties learn during the process. Black-and-white images take on grey hues, original problem definitions lose their meaning and are replaced by more fruitful problem definitions. Ambition management means that the process manager (either on his own or through the process parties) tries to match the ambitions of the parties in the process to the expectations of the environment. In other words, the environment is informed of the learning processes taking place among the parties during the process.

Here, too, the smart city is an illustrative example. Huge ambitions underlie the smart city envisaged by the Dutch Ministry of Transport. During the process of the parties involved, in which the ministry participates only indirectly, a rich picture emerges of the possibilities of a smart city. Because the participants in the process fail to make it sufficiently clear to the ministry that the ambitions need adjusting (which comes down to insufficient ambition management), the gap between expectations and reality becomes so wide that the ministry will at some stage intervene to save as much of the original ambitions as possible. This intervention is highly adhocratic and might have been far more elegant if ambitions had been properly adjusted.[70]

7.5 CONFLICTS ARE TRANSFERRED TO THE PERIPHERY OF THE PROCESS

Chapter 3 already discussed this design principle and described how and why conflicts should be kept as far away from the centre of the decision making as possible: given the conflict of interests between the parties, the conflict handling potential of the centre of the process may be limited. If there are too many conflicts at the centre, there is a risk that the substance will force out the process: the process stagnates, preventing the advantages of keeping a process going (see chapter 5) from being realized.

The organization of a process may be used to reduce conflicts. The organization will always have a number of persons who form the core of the process and persons who are rather on the periphery of the process. Suppose there is a *steering committee–project group structure*. The core of the process then comprises the members of the steering committee, representing the parties on a high level and having to take decisions together. The next skin contains the deputy members, for example. The skin that follows is the position for the project groups, preparing the steering committee's decisions. The next skin contains working groups, while the outer skin contains the parties outside the process, which, however, take or have an interest in it (see figure 7. 2).

Figure 7.2 The organizational structure of a process

The process manager may prevent conflicts in a process being sucked into the centre by framing them so that they do not always have to be solved by the parties at the centre of the process. We present a number of strategies.

- *Frame conflicts so that they have to be settled outside the process.* The most convenient situation is the one in which conflicts have to be solved by actors outside the process. Consequently, the negotiating parties do not have to give a decision about them and the conflict places no burden on their mutual relations.
- *Frame conflicts so that they have to be settled in the outer skin of the process.* If this is impossible, another option is to submit a conflict to a body outside the centre of the process. The process manager will have to prevent all conflicts having to be solved in his steering committee. In some cases, it is useful to formulate them in such a way that they have to be solved by the project groups referred to above.
- *Frame conflicts so that they can be solved on the level of deputies.* It is important for representatives with commitment power to work with deputies frequently. A framing of conflicts that allows deputies to solve most of them eases the pressure on the relations between the parties. This is because it

leaves less potential for conflict for those representatives of the parties who are responsible in the first place.

> Deputy members can play an important role in decision-making processes.
> Sparks mentions a number of these advantages in his description of the negotiating process between the South African government and the ANC:
> - they may be used to get some understanding of a party's room for negotiation;
> - they can help when relations between representatives are difficult, for example because parties have negative images of each other;
> - they form extra channels for disseminating and receiving information.[71]

- *Frame conflicts in such a way that the coalition of proponents and opponents is always different.* Of course, particularly in the last round of the process, a number of conflicts will remain that require solving by the inner skin, the members of the steering committee in the example. An important rule for the process manager here is that the forming of always the same coalitions in these conflicts should be prevented. Problems can be formulated in such a way that the coalitions are not too predictable and also change a number of times. If the coalitions were always the same, a block may form at the centre of the process, which may seriously jeopardize the process.

Is all this a sign of weak leadership and insufficient decisiveness? Here, too, the answer must be that – in a network – there is little chance of success if a command-and-control style is chosen. It is more intelligent to give the process a chance by reducing conflicts, allowing it to have a positive effect. If the process takes its course (see again chapter 5), a breeding ground may be created for decision making and even for command and control (see the following section).

7.6 COMMAND AND CONTROL IN A PROCESS: AS A DRIVER AND AS A RESULT OF THE PROCESS

Process management is an alternative for the use of command and control in decision-making processes. The reason is that command and control is unlikely to succeed in a network, in which none of the parties is hierarchically superior to the other parties.
This is not to say, however, that command and control is meaningless in a process. Command and control can play a role in process management if it is made to serve the speed of the process. This can be realized in two ways. Command and control may be a process driver (strategies 1 to 3), and, in addition, a good process is a breeding ground for command and control (strategies 4 to 6).

Strategy 1
Command and control may be a process driver because it puts pressure on the parties in a process. Suppose there is a process going on in which the dynamic of a process is effective: relations are being developed and possibilities for gain emerge. The process may be speeded up, however, if forms of command and control are used. A minister negotiating with industry about a partnership agreement in a process, but threatening to bring in unilateral rules at the same time, is likely to reach a better partnership agreement and reach it sooner than a minister opting for a process only. This is also found in the relations between states in the form of 'bulldozer diplomacy': negotiations in a process are accompanied with a show of strength to speed up the negotiating process.
An important effect of these and similar forms of command and control is that it gives parties a different perception of their gains. Gains are not only the results they can achieve by negotiating in the process, but also the prevention of the threat emanating from command and control. In the example of the negotiating minister: preventing unilateral rules may be a form of gain. For industry to cash in on these gains, it will have to be guided by the minister's wishes in one way or another. So there is every chance that this will speed up the consulting and negotiating process.

Strategy 2
Command and control can be a process driver if it is used while room for a process is offered simultaneously.

> The Board of Management of an organization with the structure and the culture of a network may unilaterally announce a merger between two divisions and simultaneously offer room to these two divisions: they can exert a strong influence on the strategy and structure of the new, merged division in a consultation process. This faces the divisions with a trade-off: resistance to the merger decision (negative energy) or making optimum use of the room offered (positive energy). If they opt for resistance, this will require an almost impossible control, since there will always be units within the division that opt for using the room offered and will therefore be difficult to control.

If the Board of Management confined itself to designing a process for achieving a merger, some of the consequences are easy to predict: reactive behaviour by the divisions, attempts to delay the process, no loyal participation in the process, et cetera. Here, too, the combination of command and control and process management may ensure a certain speed of the process.

Strategy 3
Command and control may be a driver if it is used to create among the parties a sense of urgency regarding a process. As we pointed out earlier (chapter 4), a process only has a chance of success if the parties feel such a sense of urgency.

> An often-heard complaint is that a particular initiative starts off as a project and then degenerates into a process. A design for a project is made, which helps to activate the parties. They find the design in conflict with their interests and oppose its realization. Such a development is regarded as undesirable and may be a reason to recommend that an initiative should be developed as a process rather than a project: in the process, invite the stakeholders to set up a project. There is something inevitable about the development from project to process, however: no process without a project. The project is the driver for parties to become active; if there is no project, parties will, as a rule, not be interested in committing themselves to a process. In other words, the project creates a sense of urgency among the parties: the project teaches them that only in a process can they arrive at a decision for which there is broad support.

An active use can be made of this inevitable (and, occasionally, tragic) development. The manager may present a detailed project, not in order to realize it, but as a driver for a process. This is a form of command and control.

Strategy 4
Command and control may be helpful when, in time, a critical mass of parties stands to gain by the process. This is because these parties will wish to cash in on their gains and have an interest in completing the process quickly. They will put pressure on the other parties–either consciously or unconsciously, either explicitly or implicitly–to complete the process and can use for this purpose the networks of connections that were built during the process.
This may suddenly accelerate the process. This is the reason why many processes end in a pressure cooker: a sudden acceleration of the process, because a critical mass of parties wishes to arrive at decisions. A special dynamic may occur in this pressure cooker, but this may be difficult to predict.

- Potential losers trying to block the decision making in the pressure cooker may be compensated liberally by the parties that stand to gain. This is because the latter want to reach decisions quickly and are therefore sooner inclined to give in.
- The other parties may also put huge pressure on potential losers trying to block the decision making in the pressure cooker. This changes their perception of gain: the losers also want to prevent their relations with the potential winners

deteriorating rapidly because the losers block the decision making, and will rather interpret a particular decision as gain.

- Process-based rather than substantive decisions may be taken about the subjects that are highly important to the potential losers ('further consultation about subject X will follow'; 'the parties decide to take no actions regarding subject Y without the approval of party X', et cetera). In many cases, this means that the parties continue their interaction in another process, with a new agenda, sufficiently attractive to the potential losers to help complete the current process. This is sometimes referred to as a *roof tile construction*: the last round of a process is designed so as to be the first round of a new process. Thus, various decision-making processes are coupled and overlap, causing the end of each decision-making process to be influenced by the next decision-making process. Dixit and Nalebuff formulate it as follows: "To avoid the unraveling of trust, there should be no clear final step. As long as there remains a chance of continued business, it will never be worthwhile to cheat. So when a shady character tells you this will be his last deal before retiring, be especially cautious."[72]

Strategy 5

Command and control is attractive after a process has been completed successfully. A manager who, in a process, has successfully consulted with the divisions in his organization and has reached agreements with these divisions may then present the result of these agreements as a firm and unilateral decision. There is firmness in the public performance, while negotiations take place backstage.

This public performance is important because it boosts his authority and may strengthen his position in the following rounds of negotiations.

Strategy 6

Command and control has a chance when a process has failed. If a manager asks a number of parties to arrive at decisions in a process and they are unable to do so, room is created for unilateral decision making. The reason is that the parties learn that they are unable to solve particular problems in mutual consultation. They learn that a process brings high decision-making costs, which is one of the reasons why they are sooner prepared to accept unilateral interventions. The process thus creates a breeding ground for command and control.

Now a process may fail without the parties envisaging or foreseeing it. Instead, a party may be convinced in advance that a process will fail. The following two situations may then occur.

- A party wants to implement a strategic plan and is convinced that the stakeholders will be unable to reach a decision by mutual consultation. Given this conviction, the party decides that the strategic plan must be implemented. This provokes resistance among the stakeholders, who will give up their resistance after some time and accept implementation of the plan.

- A party wants to implement a strategic plan and is convinced that the stakeholders will be unable to reach a decision in mutual consultation. It nevertheless spends time on and creates an opportunity for a process. The process fails, making the stakeholders more amenable to a unilateral decision and allowing the implementation of the strategic plan to be announced. The manager's intervention is so strong that they accept his decision.

The party concerned must decide which of the two situations brings the lowest decision-making costs. At first sight, the second situation is a waste of time, but it may be efficient to allow the parties some time, to make them learn in a process that they are unable to reach a decision.
Parties who merely resist a decision (first situation) will not move through this learning process. The question is therefore which costs are higher: those of a failed process (second situation) or those of the parties' resistance (first situation).

Notes

[64] See for example Ch. G. Field, "Building Consensus for Affordable Housing", in: Housing Policy Discutee (1997), No. 4, pp. 801-832, in which especially the Hartford case study is illustrative. Research by Parker and Wragg points in that direction. Realizing a counternetwork as distinct from the existing network strongly influences the course of the decision making. Gavin Parker and Amanda Wragg, "Networks, Agency and (De)stabilization: The Issue of Navigation on the River Wye, UK", in: Journal of Environmental Planning and Management (1999), pp. 471-487; J. Huygen, "Cultural and strategic dimensions in decision making: Techno-politics and the GBA project", in: P. 't Hart, M. Metselaar and B. Verbeek, *Political decision making*, The Hague, 1995, pp. 125-148, 136 (in Dutch); M. Rein and D. A. Schon, "Frame-reflective Policy Discourse", in: Beleidsanalysis (1986), No. 4, p. 8; R. Tolmach Lakoff, The Language War , Berkeley, 2000, p. 9.
[65] Van den Bosch and Postma (1995), p. 5.
[66] idem
[67] De Bruijn, Ten Heuvelhof and In 't Veld (1998).
[68] Kotter (1997), pp. 34-40.
[69] Weening (2001).
[70] Weening (2001).
[71] Sparks (1995), pp. 31, 78-79.
[72] Dixit and Nalebuff (1991), p. 158

8. THE PROCESS MANAGER AND THE SUBSTANCE OF DECISION MAKING

8.1 INTRODUCTION

The fourth core element of the process approach is substance: the process developed under the guidance of the process manager must be sufficiently substantive, since a process without substance is empty.
The preceding chapters already set out on a number of occasions that a decision-making process may degenerate into a process for the sake of the process. This may affect its speed (core element 3), but also its substance. A process drifting away too far from the substance is vulnerable and falls short of its original aim: it is designed to generate substantive problem definitions and problem solutions.
This chapter describes how the quality of the substance in a process can be protected. Section 8.2 examines the role of experts in a process. What is their relation to the stakeholders and how do they contribute their expertise? By way of intermezzo, section 8.3 explores the relation between strategic behaviour on the one hand and substance-driven behaviour on the other hand. Section 8.4 outlines a desirable course of the process from a substantive perspective. It introduces and operationalizes the standards of variety and selection.

8.2 UNBUNDLING AND BUNDLING EXPERTS AND STAKEHOLDERS

A process approach to decision making is used because a purely substantive approach is impossible. This is because the problems to be solved are unstructured, which precludes an unequivocal substantive solution.
The attendant risk is that the process forces out the substance: the interests are so dominant that they push aside the substantive forces in the process. For example, for the sake of consensus, parties accept the outcome of a process that is attractive to all of them, but that will not stand up against existing scientific insights. 'Anything goes': parties simply decide that a particular problem definition and problem solution are the correct ones and do not allow themselves to be corrected in any way by substantive insights or by the views of experts. Thus, an opportunistic use is made of the fact that a problem is unstructured.

> In his study into the realization of large public projects, Robert Bell finds that the building of many of these projects is started before the design is sufficiently crystallized. He argues that a design has two functions in a context of conflicting interests. There is a substantive function: a design should direct the building of a project. In addition, there is a process function: a design should serve the interests of

> the parties involved. The substantive function may suffer if this process interest is too dominant. For example, designs may be created that conflict with the laws of physics, which will obviously seriously affect the building of the project. For some parties, this is no problem from a process perspective: once it has been decided to build the project and the building has been started, there is hardly a way back in the case of many large projects. [73]

If there is an unstructured problem, the parties should, ideally, seek *negotiated knowledge*: substantive knowledge (1) accepted by the stakeholders and (2) proof against scientific criticism. If the process forces out the substance, *negotiated nonsense* tends to emerge, which fails to meet the second criterion.

In addition, there is the risk that the decision making in a process lacks innovation. Too little use is made of new insights, for example because the administrative parties participating in a process are simply unaware of them.

The substantive expert finds himself in a tense position: on the one hand, he is not the substance expert who can give the right answers to the questions asked by the stakeholders. This is because problems are unstructured, which is why different, conflicting expert views of particular problems may all be correct. On the other hand, he does have substantive expertise, ignoring which will only harm the parties. How to deal with this tension in a process?

The answer is that process agreements have to be made about the role of experts as well. These are based on two pillars.

Unbundling of roles, ...

Experts distinguish themselves from the stakeholders by their expertise, which is less strongly tied to a particular interest. This is an argument for making a clear distinction between experts and stakeholders in a process. Such unbundling implies that the expert and the interested party play different roles. The expert can advise the parties and also plays an important role as 'countervailing power' towards the stakeholders. For example, allowing experts to take a critical look at the draft outcome of a process may prevent negotiated nonsense. If there is no such unbundling of roles and no clear agreements are made on this point, there is a risk that the expert will assume the colour of one of the stakeholders. In that case, he is not the person who takes a critical look at the results or interim results of a process or indicates what innovations are possible, but someone who justifies decisions by providing substantive argumentation for them.

Unstructured problems are particularly open to this risk, because there are no unequivocal solutions for them.

... *followed by a bundling of activities*

The idea of unbundling of roles needs an important addition as table 8.1 explains.

	Unbundling	Bundling
Advantages	Expert can act as countervailing power.	Influence on decision making
Disadvantages	The authority of expertise is a problem and it is submitted at the wrong time.	Expert assumes the colour of stakeholders and perverts.

Table 8.1 The roles of expert and interest party unbundled and bundled

Unbundling experts and decision makers tends to be based on the idea that the expert or researcher discovers the facts, after which the decision maker takes a decision. Although such a distinction is outmoded from a philosophy-of-science point of view, it continues to play a role in the practice of decision making. A logical consequence of such a view is that the roles of experts and decision makers are strictly separated: first the facts, then the judgements. In the case of unstructured problems, such unbundling has two major disadvantages, however:

- The knowledge contributed by experts has no authority for the stakeholders. They may very well decline to accept the results of a study, for example, because they disagree with the data, methods or system boundaries chosen. Garbage in, garbage out, as the saying goes.
- A temporary misfit develops between the knowledge contributed by an expert and the decision-making process. The results of a study become available either too early or too late, for example. In this way, serious criticism by researchers of a particular decision by the parties cannot influence the decision making. Unbundling reinforces this mechanism. Minerva's owl leaves its nest when it's dark, says the proverb.

Research by Jasanoff corroborates this. She concludes that processes in which scientific study and decision making are strictly separated stand little chance of authoritative and consolidated decision making. There is a better chance of such decision making in processes in which science and decision making mix.[74]
This rather points in the direction of bundling between research and decision making. Bundling implies that experts have a better insight into the course of the decision making and can therefore intervene better at the right moment. Bundling also means that experts have more possibilities of dealing with parties' criticism of their analyses. Bundling implies intensive interaction, enabling experts to react properly to parties' criticism.

They know parties' views better, know the inconsistencies in these views, have more possibilities of iterations in their research, et cetera.
This produces an ambiguous picture: bundling is necessary, but involves the risk that experts take on the colour of the stakeholders; unbundling is therefore desirable, but involves the risk that experts can play no authoritative role in the decision making. What does this mean for the relation between experts and stakeholders in the process?

- On the one hand, the *roles* of experts and stakeholders should be unbundled (see above).
- On the other hand, it is necessary to bundle the *activities* of the two parties. They should interact intensively from their unbundled roles, preventing the above misfits from occurring. Unbundling prevents experts from taking on the colour of one of the other stakeholders.
- Bundling of activities may be achieved by the process agreement that the stakeholders have to submit their results or interim results to the experts at particular moments in the process and may submit these results or interim results at other moments.

Such bundling proceeding from unbundled roles has two functions: it improves the quality both of the decision making and of the knowledge contributed by experts.[75]

Improving the quality of the decision making. First, bundling implies that the quality of the decision making by the stakeholders will improve. This is because bundling will force the stakeholders to submit their views and suppositions in the meeting with experts. The experts will not indicate at that stage which proposals and views have to be chosen (since there are unstructured problems, precluding an authoritative indication), but indicate how the proposals and views submitted by the parties can stand up against scientific insights. It may be found that particular views or suppositions cannot stand the test of scientific criticism.

> The Orange County Landfill Selection Committee has to decide on the location of a landfill. It can choose from seventeen potential locations, which can be compared for sixteen variables.
> The Committee commissions a study, expecting it to give an objective answer to the question where the landfill should be located. The study produces a different outcome, however. It shows what judgements can be regarded objectively and where there is room for parties to negotiate. The outcome of the study is that only four of the sixteen variables are relevant when comparing the locations: groundwater, surface water, covering and isolation. The other twelve variables do not constitute differences between the locations.

> The study also shows the score of the different locations for the four variables. A particular type of judgement, for example, is that location X pollutes the groundwater in particular.
> The study can thus make the decision making more substantive. Parties invoking one of the twelve non-discriminating variables in the process would seem to have a weak case in the decision-making process. The study also shows where there is room for negotiations because unequivocal judgements are impossible.[76]

The involvement of decision makers reduces the risk of a temporary misfit between the decision making and research. The decision makers are aware of the current research and also slightly committed to it because of the bundling. Conversely, experts are more able to keep in touch with the decision making and thus know better when there is momentum to submit substantive expertise.

Improving the quality of the knowledge submitted. Second, bundling improves the quality of research, since a critical stance of the stakeholders on the study and its results will demonstrate what the underlying values are, what data, methods and system boundaries are used, which results are robust and authoritative and which results fail to meet these criteria.

> A study by an independent firm of consulting engineers shows that the polycarbonate bottle rates low against other packages. A proponent of the polycarbonate bottle takes a very critical look at the suppositions used. He makes these suppositions more explicit and manages to convince the other parties that they are incorrect. This improves the quality of the study into the environmental effects of the polycarbonate bottle.
> The fact that every package has its own proponents in the process guarantees that the analyses of the other packages will be considered just as critically.[77]

Research and decision making: parallel connection and proper bundling. What are the implications of the above for the planning of research activities in a decision-making process? This question is particularly relevant for processes that are highly information-intensive and of which it is clear from the start that a great deal of research has to be done. For example, the decision making about large infrastructural projects will nearly always bring a great deal of research: into the environmental effects, the economic effects, the consequences for safety, noise nuisance, et cetera. The design question is: how should the research be planned in relation to the decision making?
After what was said above, it will be clear that a serial planning (research first, followed by decision making) is undesirable. Parallel planning is preferable. The research and the decision making should be bundled. There are three tasks facing the process manager:

- he should keep the roles of experts and stakeholders separate;
- he should ensure proper bundling of the activities;
- he should ensure a parallel connection between the research process and the decision-making process.

Once the new tasks have been performed, the process manager can be an interface between experts and stakeholders. He can use the knowledge of experts to make the process more substantive. We will list six situations in which the use of experts makes the process more substantive.

Options for improvement
A choice must be made between two options: the existing option A and the new option B. The decision makers have to decide which option performs better. Research shows that this is option B. The decision makers then want to know whether the performance of option A and option B can be improved. Researchers extrapolate the improvement options of A and B. Research shows that option A has more options for improvement than option B.
The options for improvement might not have been extrapolated without experts playing an active role. The decision makers only become interested in the improvement options of A after B's options are found to be better. Had the research and the decision making been connected in series, the calculation of the improvement option might not have been available.

Standardization
A choice must be made from six options: options A to F. Experts indicate that options A to D do considerably worse than E and F. The decision makers eventually select options E and F and a choice has to be made from these options. The options should be compared, but are difficult to compare. Researchers conduct a standardization: the options are made comparable using a particular method. Option E is found to be better than option F.
A standardization of A to D might also have taken place without experts playing an active role, or there might have been no standardization, as a result of which the choice from option E and option F would have been less well-founded from a substantive point of view.

Sensitivity analysis
In a decision-making process, a number of stakeholders are against choosing option A: option A is bad, because option B does well if there is faster economic growth. Researchers then conduct a sensitivity analysis: what is the performance of option A if there is faster economic growth?

It would not have been conducted without experts playing an active role; without a parallel connection, too many sensitivity analyses or different sensitivity analyses might have been made.

Enriching the decision making

In a decision making process, five technical options are available: options A to E. The decision makers find that a number of combinations are possible: A+B, C+E or D.

Research shows that there is a fourth possibility: A+B+C. This possibility might not have been submitted without experts playing an active role, and the decision making would have been less rich.

> The process about packages started with the simple question whether the disposable or the non-disposable package is better for the environment. Such a dichotomy easily leads to discord and does not lend itself to coupling problems to solutions, since the parties take the view that a choice has to be made between the disposable or the non-disposable package.
>
> The contribution of experts caused the parties to give a more finely tuned description of the problems after the process, as a result of which far more creative couplings of problems to solutions are possible. A number of examples from the final report by the parties:
>
> - The disposable–non-disposable dichotomy is wrong. There are 'hybrid packages' as well (refill packages, for example), which are partly disposable and partly non-disposable and have a good environmental performance.
> - Disposable additions (caps, clips, et cetera) count heavily in establishing the environmental impact of non-disposable packages.
> - The weight of the packages is very important; return transport of non-disposable packages within the Netherlands is less relevant.
> - As a rule, non-disposable packages are less optimized than disposable packages.[78]

New options submitted by experts may have a positive effect on the process, since they mean new possibilities of negotiating and reaching a decision.

Room in the decision making

Six options are available in a decision-making process. One dominant party reports that option A is technically unfeasible and submits an existing study to prove this. The other parties take a critical look at this study and then ask the researchers a number of questions. As a result of this interaction, the researchers have to admit that option A is feasible under certain conditions. They also have to admit that option E is technically unfeasible with the assumptions used.

This room might not have been created without experts playing an active role.

Relieving the agenda
Six options are available in a decision making process. The parties involved have to negotiate about them. After some time, the parties agree about the variables for which the options will be assessed. Research shows that options A and B do far worse than the other options. The parties then decide to disregard these options. Without an active role of experts, pressure would not have been taken off the agenda.

> An interesting form of bundling research and decision-making is the way the study into environmental effects of large physical work is embedded in the decision making process about these projects in the Netherlands.
> The Environmental Management Act makes an Environmental Impact Statement compulsory for particular projects. In short, the rule obliges the initiator of a project to study its environmental effects and report on them in an environmental impact report. He must also develop a number of alternatives to his proposal and describe them in the same report, including the environmental effects of these alternatives.
> Although the rule is not popular with initiators (nor with the competent authority, in most cases), the environmental impact statement, reflecting the research conducted, is found to have a strong effect on the thinking and acting of the actors involved.
> The Environmental Impact Statement involves the decision makers in the research. It is the initiator who proposes the alternatives and describes the environmental effects. The competent authority fixes the requirements the research has to satisfy. Usually, a specialized research institute actually conducts the study, thoroughly supervised, of course, by the commissioning party or, as the case may be, by the initiator. The results are tested for their scientific quality by a committee of independent experts. Usually, this committee also believes that too much criticism might be unhelpful to the position and the effect of the Environmental Impact Statement.
> In short, research and decision making are professionally separated, but actually bundled. The decision makers leave room to formulate a problem definition, to place accents, to emphasize particular conclusions, and there is intensive contact between researchers and decision makers.[79]

8.3 INTERMEZZO: STRATEGIC BEHAVIOUR OR SOUND AND SUBSTANTIVE BEHAVIOUR?

Parties behave strategically in complex decision-making processes: they do not disclose their views, for example, because they want to keep their options open or rather adopt an extreme view to strengthen their negotiating position.
Such strategic behaviour would seem to harm the substance of a process. Parties should not permit themselves to be guided by strategic considerations (or, which is worse, opportunistic considerations) in determining their position, but by

substantive ones. In this section, we will pay attention – in the form of an intermezzo – to the distinction between strategic behaviour and behaviour driven by substantive considerations. We will argue that this distinction is problematic.[80]

A distinction between substantive behaviour and strategically inspired behaviour is part of a particular way of thinking about process management, which may be summarized as follows:

- Good substance is created in a good process.
- A good process implies, among other things, that problem definitions and problem solutions should not be allowed to become fixed too soon. The participants in a process will first have to diverge (contemplate a large number of problem definitions and problem solutions) and should only be allowed to converge at a later stage (choosing one or more problems and solutions). The idea is that the quality of the problem definition and problem solution is good as long as they emerge from a wide variety.
- This movement from divergence to convergence tends to be hampered by parties' interests. Interests block parties' open-mindedness, necessary to diverge.
- Inherent to this is that parties will behave strategically when driven by interests. For example, they have hidden agendas and will thereby disturb the process of creating good substance. Processes of power will thus corrupt the substantive debate.
- A process of divergence and convergence can only develop in a kind of 'power-free room'.
- Process management is aimed at creating such room. The parties agree, for example, that they will hold a businesslike and open debate, that positions will be equal in the substantive debate, that only the power of arguments will play a role, that they will not play power games, et cetera.
- Process management thus comprises two core elements: guiding the substantive debate (with the help of all sorts of communication techniques) and – prior to this – making agreements with parties about not using strategic behaviour. This allows the process to take place in the power-free room referred to above.

Many forms of interactive policy development are based on assumptions of this kind. Substance is good, strategic behaviour is bad and one of the aims of interactive policy development is not to have substance corrupted by strategic behaviour.[81] Two objections can be raised against these approaches: they are naive and they are based on an unjustified distinction between power and substance.

Substance is superior to strategic behaviour.
This approach is naive: the parties involved will behave strategically as soon as a process deals with decisions to be taken. After all, parties are out to advance their interest (power) and have a perspective of problems and solutions just as legitimate as that of other parties (substance).

In many cases, such strategic behaviour is legitimate as well.
Different parties may have different, legitimate perspectives of problems and solutions when there are unstructured problems.
The perspective they develop may be shaped by the parties' position and their self-interest. When propagating their perspective, they do so partly by substantive argumentation, partly by trying to strengthen their position in the network (entering into alliances, biding their time, et cetera). In other words, they do so by exercising power.
This exercise of power is not aimed at corrupting substance, but at allowing a legitimate perspective to play a role in the decision making. In a decision-making process, different parties with different legitimate problem definitions and problem solutions compete, with the aim of exercising maximum influence on the envisaged decision. Those who do not join this game may see their own perspective go to the wall in the decision making. This may affect the quality of the eventual decision. In other words, the quality of a decision partly depends on strategic behaviour.

Distinction between substance and strategic behaviour debatable
It is clear from the above that the distinction between substance and power is debatable. It is true, no doubt: power and exercising power may corrupt the substantive debate. In a network in which problems are unstructured, the distinction between substance and power is far from unequivocal, however.

- There is a kind of strategy to promote substantive views in a decision-making process (see above).
- Substantive argumentation can be used strategically, either consciously or unconsciously.
- This gives rise to a dynamic that is difficult to unravel. If a party in a process changes a view, is this a substantive enrichment of that view or an opportunistic change, prompted by considerations of power?

Occasionally, the distinction is difficult to make in practice as well, and parties may hold different views about it. One of the experiences with process management is that party X blames party Y for strategic behaviour, enraging party Y: its behaviour was not strategic, but inspired by noble, substantive motives only.[82]

If there is no unequivocal substantive solution, the legitimacy of perspectives and interests increases, causing the legitimacy of actions to advance this perspective or interest to increase as well.
So we fail to see the objections to strategic behaviour if parties have a legitimate interest or a legitimate perspective, as long as they stick to particular rules of the game in their strategic behaviour. The same goes for objections to hidden agendas, for example. What is wrong with a hidden agenda, if parties know that strategic behaviour is inevitable and may even be legitimate? Hidden agendas are part of the game in a negotiating environment and need not harm trust between parties if they play the game by the rules.
If parties have to take a decision, strategic behaviour is a fact and it is legitimate. A process design will then have to be shaped so as to leave room for this strategic behaviour. Something similar applies to process management. This means that process design and process management must largely accommodate parties' natural behaviour: the rules of the game for negotiations are stated with respect for the fact that parties will want to behave strategically. What is more, the question what is strategic behaviour and what is not is hardly interesting if a process design facilitates consultation and negotiation.

Process design: creates power-free room	**Process design: facilitates negotiations**
Unbundling of power and substance	Power and substance are inseparable, both factually and normatively.
Creates power-free room.	Creates framework for negotiations.
Offers rules of the game for substantive enrichment.	Offers rules of the game for decision making.
Strategic behaviour is disturbing and therefore undesirable.	Strategic behaviour is natural and therefore permitted.

Table 8.2 Two types of process design

Such a process design is the mirror image of the one seeking power-free room (see table 8.2). In the latter, strategic behaviour is excluded, and then a substantive debate is held. The opposite of this is a form of process design that regards decision making as a negotiating process and therefore accommodates strategic behaviour. In this respect, too, a process design is contingent and the paradox of change presents itself once again: Those who are consistent with the existing can change.
If openness is offered and strategic behaviour is banned in a process design, everyone will suspect that there is strategic behaviour, which will generate conflict, since it is contrary to the agreements. If the distinction between

substantive argumentation and strategic behaviour is vague, there is a reasonable chance that such suspicions will arise.
If strategic behaviour is permitted in a process design–allowing parties to show their natural behaviour–such suspicions are far less significant. Trust between parties may grow during the process, which may moderate their strategic behaviour.

8.4 THE PROCESS FROM SUBSTANTIVE VARIETY TO SELECTION

Unbundling and bundling therefore enable experts to contribute to the substantive quality of decision making. The next question is what the most desirable process is from the perspective of the substance and the quality of decision making.
The question about the *substantive* quality of the decision making again receives a *process-based* answer. There is quality if the decision making process is a process of *variety and selection*. The greater the variety of options discussed, the better the quality of a decision-making process. The quality of the decision-making process will suffer if a particular option is not included in the decision making. Variety is important for the following reasons:[83]

- The greater the variety from which the eventual option results, the more authoritative the option will be. Its quality will be less debatable than if it defeats a limited number of other options.
- The actors involved are offered maximum chances of learning processes. The actors participating in the process take note of the options, reflecting on their strengths and weaknesses in interaction.
- In the discussions about them, possibilities will be explored of improving options and making them acceptable to a maximum number of actors. This improves the quality of the options.

At some stage in a decision-making process, the transition will be made from generating variety to selecting the best option or options. This selection will have to lead to consolidation: the option or options selected will have to stand up to criticism for some time.

- The first condition for a consolidated selection is that a high variety of options was contemplated (see above).
- The second condition is that there must be a link with the variety of options contemplated. There will be no authoritative selection, for example, if options are selected that were not contemplated in the variety phase.
- The third condition is that the option selected has the support of the parties involved.

A process from variety to selection also has advantages from a process perspective. If there is sufficient variety at the start of the process, it is more difficult for the parties to submit new ideas in the selection phase. In other words, it simplifies the consolidation of the decision making.

The transition from variety to selection

The criteria of variety and selection are hardly operational as yet. How serious should an option be in order to be contemplated? When should the options be known? Who decides what option will be taken seriously? What actually is 'contemplating'? These questions have to be operationalized if the requirement of variety and selection is to provide a firm basis.

A process manager who in a process faces the question about the chance of new, important options arising in the process, can form an impression by checking whether the decision-making process is continuing to produce learning processes among the parties involved. This operationalization therefore runs as follows: *It is fruitful to continue the variety-generating phase as long as actors participating in the process continue to learn.*

Two forms of learning may be distinguished: cognitive learning and social learning. Cognitive learning involves the question whether parties still produce new facts, views, values, arguments, patterns of thinking, et cetera,. As long as the process manager concludes that this is still the case, there is a chance of new variety being created. There is no longer any cognitive learning if the parties go on reproducing facts, views, et cetera.

Social learning involves the question whether parties still establish new relations and interactions, as these may bring new insights and therefore new variety. All this means that many processes from variety to selection do not proceed smoothly, but look chaotic and messy. This may seem to point to a bad process, but it may rather be a sign of the quality of a process.[84] (Also see figure 8.1.) A process is ripe for selection if there is cognitive and social stabilization.

Selection takes place at moments 1, 2 and 5. The figure also shows that iterations occur in the process from variety to selection: moments 3 and 4. Occasional selection has taken place, but participants reverse it, for example because they have established new relations. Of course, the process will be jeopardized if this is done too often. This is another argument for not ending the variety-generating phase too early.

Here, too, it is easier for parties to participate wholeheartedly in the process if they are offered room for iterations.

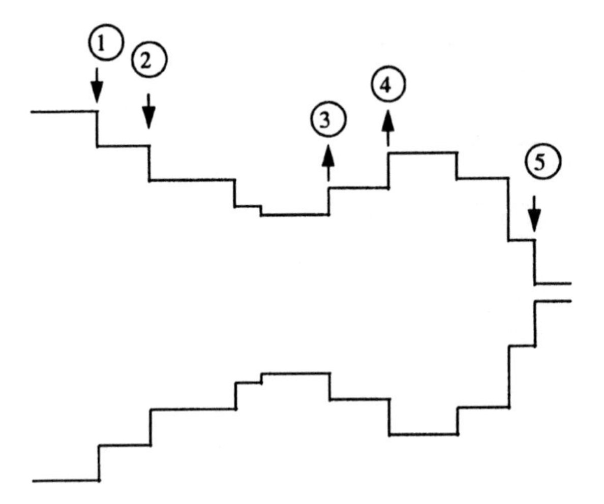

Figure 8. 1 A process from variety to selection

The process manager should then do his utmost to prevent this room being used. Table 8.3 summarizes the argument of the preceding sections. Two arrangements may safeguard the substantive quality of decision making in a process: (1) unbundling the roles and bundling the activities of experts and stakeholders, and (2) a process from variety to selection. Table 8.3 shows the consequences for the substance and the quality of decision making, and also shows that these arrangements also have positive effects from a process perspective.

The substantive expertise of the process manager

The above shows that the balance between substance and process is the process manager's major concern. He must prevent the substance forcing out the process or the process forcing out the substance. This finally prompts the question of the substantive expertise the process manager should have.

	Consequences for substance and quality of the decision making	**Consequences for the process**
Experts and stakeholders: unbundling of roles, bundling of activities	A sense of the capriciousness of decision-making process More authority of research findings Expert is not influenced by parties' interests; no 'changing of colour'	Efficiency: particular options excluded from the decision making. Room; new options may simplify decision making.
Process from variety to selection	All possible options have been contemplated. Learning processes More authoritative selection	Consolidated decision making: evading the decision making becomes more difficult.

Table 8.3: Summary of the argument

Earlier, we described the risks attached to a model in which the process manager also acts as a substance expert. The process manager's interference with the substantive course of events may then be so strong as to put pressure on the process aspects.

It may then be asked, however, what substantive knowledge the process manager should have himself. The process manager must have substantive knowledge even if he does not play the role of substance expert. If not, he risks being not taken seriously by the parties. A process manager allowing himself to be guided by substantive knowledge runs the risk that the substance will force out the process, thus harming his position as well.

This tension may be solved by distinguishing between substantive knowledge of the first, second and third order:

- Substantive knowledge of the first order means that a process manager is able to ask the right questions;
- substantive knowledge of the second order means that the process manager is able to judge the answers to these questions adequately;
- substantive knowledge of the third order means that the process manager himself is, or would be, able to answer these questions.

A process manager should at any rate have substantive knowledge of the first order. A good process manager will usually manage to acquire substantive knowledge of the second order during the process.

The risk of substantive knowledge of the third order is that the process manager will interference too much with the substance of the process, thus harming his position.

Notes

[73] Robert Bell, *The Worst Practice* , Rotterdam, 1998 (in Dutch: De bodemloze put)
[74] Sparks, 1995, p. 231. Also see, for example: Yoichi Tanaka and Ryo Hirasawa, "Features of Policy-making Processes in Japan's Council for Science and Technology", in: *Research Policy* (1996), pp. 999-1011.
[75] See for example: T. Garvin and J. Eyles, "The Sun Safety Metanarrative: Translating Science into Public Health Discourse", in: *Policy Sciences* (1997), pp. 47-70.
[76] Miranda et al. (1996).
[77] De Bruijn, Ten Heuvelhof and In 't Veld (1998).
[78] De Bruijn, Ten Heuvelhof and In 't Veld (1998).
[79] E. F. ten Heuvelhof and C. Nauta, "Environmental Impact; The Effects of Environmental Impact Assessment in the Netherlands, in: *Project Appraisal* (1997), vol. 12, No. 1, pp. 25-30.
[80] These passages are derived from: De Bruijn (2000).
[81] See for example the list of references in: Edelenbos (2000).
[82] De Bruijn, Ten Heuvelhof and In 't Veld (1998).
[83] See for example: W. M. de Jong, Institutional Transplantation: How to Adopt Good Transport Infrastructure Decision Making Ideas from Other Countries?, Delft, 1999.
[84] See for example: Peter W. Salsich, "Grassroots Consensus Building and Collaborative Planning", in: *Festschrift* (2000), pp. 709-740.

EPILOGUE

Processes occupy a prominent place in this book. In the book's language: processes are designed and processes are managed. We realize that the use of heavy concepts like 'designing' and 'managing' can easily create the wrong associations and we would like to put them into perspective in this epilogue.

Using a concept like 'designing' suggests that there is nothing prior to the design. The design is then taken to have created something that was non-existent previously. Something similar applies to 'management'. It might suggest that there is a certain order thanks to the management activity during the process. Nothing would have happened had there been no management, or the process would at least have taken a chaotic course.

Both suggestions need some qualification. The first qualification concerns the idea that the process architect designs a process that would be non-existent without his creative activity. It should be emphasized that processes always exist, even without a process design. Parties will always interact and negotiate, irrespective of a design. Such processes have always been there and will always be there, whether or not there is a process architect, and a good thing too. It is remarkable, however, that too little attention is paid to processes. Proposals for change and improvement are nearly always structural rather than process-based. Proposals for improvements nearly always focus on structure. This goes for changes in domestic administration, for mergers in industry, for reorganizations and for changes of systems like those in education, health care, social security, et cetera. The idea implicit here is that processes are bound to adjust to the newly created structures. The question that prompts itself here is whether this adjustment will be so easy and will happen so fast and, more fundamentally: it is not sensible to moderate the discussion about structures and rather focus our attention on improving the quality of processes, also in the debate? More headway might be made with less effort if the change focuses on process changes right away.

A sensible process architect realizes that processes already exist when he starts. He also knows that the process will continue after 'his' process; there is a process after 'his' process. Design means that he starts from these processes, using them. He uses current processes. He protects and accentuates what is good about them. He addresses their weak elements, however, weakening their effect or position in the process.

It should also be remembered that the position of the process architect is less autonomous than that of his colleague designers elsewhere. After all, the process design is not so much the result of a cerebral-and-creative-oriented design process, but first and foremost the resultant of a negotiating process between the stakeholders, facilitated and nourished by the process designer's ideas.

The second qualification concerns the management of processes. It should be remembered that innumerable processes pass off satisfactorily even without an explicitly appointed process manager. Evidently, many managers are able to safeguard a proper and satisfactory process. They know the implicit rules and are able to contribute to the development of these rules. They have latent skills and insights that are indispensable to guide the many parties through complex processes and reach decisions with which the parties are satisfied. To achieve this, they use their knowledge of people and organizations, coupling it fruitfully to a highly developed, delicate sense of the course of events in processes of this kind. Process management aims to make these skills explicit and develop them.

Processes have always been there, and merely accentuating and using them constitutes no fundamental change in existing administrative practice. Consequently, working out the process design and the process management will not always, and by definition, be something special.
Why then is there so much attention for process management? The reason is that process management may not be a revolution in administrative practice, but – strangely enough – an accent on process management in the practice of proposals for improvement would be a revolution. Changes and improvements immediately following processes may be successful because they are in line with what good managers would do anyway!

REFERENCES

Arnstein, S.R., "Eight Rungs on the Ladder of Citizen Participation", in: S.C. Edgar and B.A. B.A. Passet (ed.), Citizen Participation: Effecting Community Change, New York, 1971, pp. 69-91.

Axelrod, R., The Evolution of Cooperation, 1984.

Baren, N. van, Plan-hierarchical solutions: A source of social resistance, Amsterdam, 2001 (in Dutch).

Bell, Robert, Worst practice, Rotterdam, 1998 (in Dutch: De bodemloze put).

Boddy, D. and N. Gunson, Organisations in The Network Age, London, 1996.

Bohman, J., Public Deliberation: Pluralism, Complexity and Democracy, Cambridge, Mass., 1996.

Bosch, van den F.A.J and S. Postma, Strategic Stakeholder Management: A Description of the Decision-making Process of a Mega-investment Project at Europe's Biggest Oil Refinery: Shell Nederland Raffinaderij BV in Rotterdam, Erasmus University/Rotterdam School of Management, Management Reports Series No. 242, Rotterdam 1995.

Boston, J. (ed.), The State under Contract, Wellington, 1995.

Bruijn, J.A. de, "From steering to process", in: R. J. in 't Veld (ed.), Steering illusion & disillusionment, Utrecht, 1999, pp. 52-68 (in Dutch).

Bruijn, J.A. de, Processes of change, Utrecht, 2000 (in Dutch).

Bruijn, J.A. de, R. van Duin and M. A. J. Huijbregts, in: J. Guinée et al. , LCA, An Operational Guide to the ISO standards, Dordrecht, 2002.

Bruijn, J.A. de, L. Geut, M.B. Kort et al.: Procedures for decision-making about a national airport. The Hague, 1999. Report commissioned by the Ministry of Transport (in Dutch).

Bruijn, J.A. de and E.F. ten Heuvelhof, A process design for a study into travel behaviour: Memorandum of advice commissioned by the Ministry of Education, Dutch Rail passengers, regional transport and VSV+, TPM, Delft, 2000 (in Dutch);

Bruijn, J.A. de and E.F. ten Heuvelhof, Rules of the game for the negotiations between the Ministry of Education and the public transport companies about a new season ticket for students, Delft, 2000 (in Dutch).

Bruijn, J.A. de and E.F. ten Heuvelhof, A process approach for a product-based environmental policy, Delft, 1999 report commissioned by VNO-NCW (in Dutch).

Bruijn, J.A. de and E.F. ten Heuvelhof, A quality system for policy and policy processes, Delft, 1999. Report commissioned by the Ministry of the Interior (in Dutch).

Bruijn, J.A. de and E.F. ten Heuvelhof, Package Covenant II: Clustering and Process of Reporting Protocol. Report for the Ministry of Housing,

Environment and Spatial Planning and the Dutch Packaging Industry, Delft, 1999.
Bruijn, J.A. de and E.F. ten Heuvelhof, A design for a protocol for testing ALARA efforts, 1997, commissioned by the Branded Goods Foundation (in Dutch).
Bruijn, J.A. de and E.F. ten Heuvelhof, The implementation of the Malta Treaty: A design for process-based governance, Delft, 1996 (in Dutch).
Bruijn, J.A. de and E.F. ten Heuvelhof, A Design for the Decision-Making Process about Maasvlakte II, Delft, 1995 (in Dutch).
Bruijn, J.A. de and E. F. ten Heuvelhof, Networks and Decision Making, Utrecht, 2000.
Bruijn, J.A. de, E.F. ten Heuvelhof and R.J. in 't Veld, Process management: Decision making about the environmental and economic aspects of packages for consumer products, Delft, 1998 (in Dutch).
Bruijn, J.A. de, E.F. ten Heuvelhof and H.I.M. de Vlaam, Interconnection disputes, in: ITeR series (1997), No. 8, pp. 165-265, Alphen a/d Rijn.
Bruijn, J.A. de, E.F. ten Heuvelhof and H.I.M. de Vlaam, "Interconnection Disputes and the Role of the Government between Substances and Process, in: Communications & Strategies (1999), pp. 295-317.
Bruijn, J.A. de, M. Kuit and E.F. ten Heuvelhof, Sport 7, Rotterdam, 1999.
Bruijn, J.A. de, R.J. in ' t Veld et all, Procedures for the project management of Mainport Rotterdam, The Hague, 1999 (in Dutch).
Bruning, A.J.F., Large projects in the Netherlands: An analysis of the length of 20 decision-making processes: Preparatory study for the Netherlands Scientific Council for Government Policy, The Hague, 1994 (in Dutch).
Buchanan, D.U. and A. Huczyncky, Organizational Behavior: An Introductory Text, Hertfordshire, 1997.
Burger, J., et all, "Science, Policy Stakeholders, and Fish Consumption Authorities: Developing a Fish Fact Sheet for the Savannah River", in: Environmental Management (2001), No. 4, pp. 501-514.
Cameron, K.S., "Effectiveness as Paradox: Consensus and Conflict in Conceptions of Organizational Effectiveness", in: Management Science (1986), pp. 539-555.
Chavannes, Marc, The sluggish state, Amsterdam, 1994 (in Dutch).
Chisholm, D., Coordination without Hierarchy, Berkeley, 1989.
Collins, J.C. and J.I. Porras, Built to Last: Successful Habits of Visionary Companies, New York , 1997.
Cohen, M.D., J.G. March and J.P. Olsen, "A Garbage Can Model of Organizational Choice", in: Administrative Science Quarterly (1972), pp. 1-25.
Cohen Commission, Market and government, The Hague, 1997 (in Dutch).
Crozier, N. and E. Friedberg, Actors and Systems, London, 1977.

Directorate General for Public Works and Water Management North-Holland, Handbook for the open-plan process, s. l. , 1996 (in Dutch).
Dixit, A. and B.J Nalebuff, Thinking Strategically: The Competitive Edge in Business Politics and Every Day Lives, New York, 1991.
Donk, W. van den, The arena in tables, Tilburg, 1998 (in Dutch).
Douglas, M. and A. Wildavsky, Risk and Culture, Los Angeles, 1982.
Dunn, W.N., Public Policy Analysis: An Introduction, Englewood Cliffs, 1981.
Edelenbos, J., Process in form: Overseeing the process of interactive policy making about local spatial projects, Delft, 2000 (in Dutch).
Eden, C. and F. Ackermann, Making Strategy, London, 1998, p. 4.
Egan, G., Working the Shadows Side: A Guide to 'Positive behind the Scenes' Management, San Francisco, 1994.
Ehrmann, J.R. and B.L. Stinson, "Joint Fact-Finding and the Use of Technical Experts", in: Susskind, McKearnan and Thomas-Larner (1999), pp. 375-401.
Field, Ch. G., "Building Consensus for Affordable Housing", in: Housing Policy Discutee (1997), No. 4, pp. 801-832.
Financieel Dagblad, 10 November 1998.
Fischer, F., Citizens, Experts and the Environment, London, 2000.
Fisher, R. and W. Ury, Getting to Yes: Negotiating Agreement without Giving in, Boston, 1981.
Garvin, T. and J. Eyles, "The Sun Safety Metanarrative: Translating Science into Public Health Discourse", in: Policy Sciences (1997), pp. 47-70.
Giddens, A., Beyond Left and Right: The Future of Radical Politics, Stanford, 1994.
Graaf, H. van de and R. Hoppe, Policy and politics, Muiderberg, 1989 (in Dutch).
Guba, E.G. and Y.S. Lincoln, Fourth Generation Evaluation, Newbury Park, 1989.
Healey, P., "Collaborative Planning in a Stakeholder Society", in: Town Planning Review (1998), pp. 1-21.
Heuvelhof, E.F. ten, Standards of behaviour for governments in horizontal structures, The Hague, 1993 (in Dutch).
Heuvelhof, E.F. ten and C. Nauta, "Environmental Impact; The Effects of Environmental Impact Assessment in the Netherlands, in: Project Appraisal (1997), vol. 12, No. 1, pp. 25-30.
Hisschemoller, M., Democracy of problems, Amsterdam, 1993 (in Dutch).
Huygen, J., "Cultural and strategic dimensions in decision making: Techno-politics and the GBA project", in: P. 't Hart, M. Metselaar and B. Verbeek, Political decision making, The Hague, 1995, pp. 125-148, 136 (in Dutch).
Innes, J.E., Planning Through Consensus Building: A New View of the Comprehensive Planning Ideal, in: Journal of American Policy Analysis (1996), vol. 62, No. 4, pp. 460-472.

Jasanoff, S., The Fifth Branch: Science Advices as Policy Managers, Boston, 1990; Science and Public Policy (1999), vol. 26, No. 3, theme issue about scientific expertise and political responsibility.

Jong, W.M. de, Institutional Transplantation: How to Adopt Good Transport Infrastructure Decision Making Ideas from Other Countries?, Delft, 1999.

Jordan A.G., "Sub-governments Policy Communities and Networks", in: Journal of Theoretical Politics (1990), pp. 319-338.

Jordan A.G. and K. Schubert, "A Preliminary Ordering of Policy or Network Labors", in: European Journal of Political Research (1992), pp. 7-27.

Kenis, P. and V. Schneider, "Policy Networks and Policy Analysis: Scrutinizing a New Analytical Toolbox", in: B. Marin and R. Mayntz (ed.), Policy Networks, Empirical Evidence and Theoretical Considerations, Frankfurt am Main, 1991, pp. 26-59, 34 ff.

Kheel, Ph.W. and W.L. Lurie, The Keys to Conflict Resolution, Proven Methods of Settling Disputes Voluntarily, 1999.

Klijn, E.H. and J.F.M. Koppenjan, "Interactive Decision Making and Representative Democracy: Institutional Collisions and Solutions", in: O. van Heffen et al., Governance in Modern Society, Kluwer, 2000, pp. 109-134..

Kotter, John P., "Leading Change: Why Transformation Efforts Fail", in: Engineering Management Review (1997).

Lakoff, R. Tolmach, The Language War, Berkeley, 2000, p. 9.

Lambers, C., D.A. Lubach and M. Scheltema, Speeding up legal procedures for large projects: Preparatory study for the Netherlands Scientific Council for Government Policy, The Hague, 1994 (in Dutch).

Maidique, M.A. and R.A. Hayes, "The Art of High Technology Management", in: Sloane Management Review (1984), No. 25, pp. 19-31.

Mayer, I.S., Debating Technologies. A Methodological Contribution to the Design and Evaluation of Participatory Policy Analysis, Tilburg, 1997.

Miranda, M.L. et all, "Informing Policymakers and the Public in Landfill Siting Processes, in: Technical Expertise and Public Decisions, Institute of Electrical and Electronic Engineers, Princeton, 1996.

Moore, C.W., The Mediation in Process: Practical Strategies for Resolving Conflict, 1996.

NRC Handelsblad, 30-6-1994.

Parker, Gavin and Amanda Wragg, "Networks, Agency and (De)stabilization: The Issue of Navigation on the River Wye, UK", in: Journal of Environmental Planning and Management (1999), pp. 471-487.

Passet (ed.), Citizen Participation: Effecting Community Change, New York, 1971, pp. 69-91.

Perrow, C., Normal Accidents, Princeton, 1984.

Quinn, R.E., Beyond Rational Management, San Francisco, 1998.

Rein, M. and D. A. Schon, "Frame-reflective Policy Discourse", in: Policy analysis (1986), No. 4, p. 8.
Riker, H., The Art of Political Manipulation, 1986.
Salsich, Peter W., "Grassroots Consensus Building and Collaborative Planning", in: Festschrift (2000), pp. 709-740.
Schelling, Th. C., The Strategy of Conflict, London, 1960.
Senge, P., The Fifth Discipline: The Art and Practice of the Learning Organization, London, 1992.
Simon, H.A., "From Substantive to Procedural Rationality", in: S. Latsis (ed), Method and Appraisal in Economics, Cambridge, pp. 129-148.
Sparks, A., Tomorrow is Another Country: The Inside Story of South Africa' s Negotiated Revolution, London, 1995, p.180.
Stern, P.C. and H.V. Fineberg (ed.), Understanding Risk Informing Decisions in the Democratic Society, Washington, 1996.
Susskind, L., S. McKearnan and J. Thomas-Larmer (ed.), The Consensus Building Handbook, Thousand Oaks, 1999.
Tanaka, Yoichi and Ryo Hirasawa, "Features of Policy-making Processes in Japan's Council for Science and Technology", in: Research Policy (1996), pp. 999-1011.
Termeer, C.J.A.M., Dynamics and inertia surrounding manure policy, The Hague, 1993 (in Dutch).
Twist, M.J.W. van, J. Edelenbos and M. van der Broek: "The courage to think in dilemmas", in: Management en Organisatie (1998), Volume 52, No. 5, pp. 7-23, 9, Alphen a/d Rijn (in Dutch).
Teisman, G.R., Complex decision making: a pluricentric perspective of decision making about spatial investments, The Hague, 1992 (in Dutch).
Teisman G.R., Steering through creative competition, Nijmegen, 1997; Mobilising space through cooperative management, Rotterdam, 2001 (in Dutch).
Vaughan, D., The Challenger Launch Decision: Risky Technology, Culture, and Deviance at NASA, Chicago, 1996.
Veld, R.J. in 't, Playing with fire, The Hague, 1995 (in Dutch).
Veld, R.J. in 't, Northern Lights, The Hague, 1997 (in Dutch).
Veld, R.J. in 't, "The steering illusion of the purple coalition government", in: Frans Becker et al. (red.), Seven years of purple coalitions, Amsterdam, 2001, pp. 164-196 (in Dutch).
Veld, R.J. in 't, Half a loaf is better...? An essay about the government on the market, The Hague, 2001 (in Dutch).
Veld, R. J. in 't, J.A. de Bruijn, E.F. ten Heuvelhof et al. A process standard for the implementation of the Packaging Covenant, Rotterdam, 1992 (in Dutch).

Veld, R.J. in 't, J.A. de Bruijn, E.F. ten Heuvelhof et al., Recommendations for a Process Standard concerning the Environmental and Feasibility Analysis as laid out in the Dutch Covenant for Packaging, Rotterdam, 1992.
Weening, H.M., The Flywheel Long Gone? An evaluation of the course of and the approach to the process surrounding the Smart City, Delft, 2001 (in Dutch).
Willke, H., Systemtheorie: Eine Einfuehrung in die Grundprobleme der Theorie Sozialer Systeme, Stuttgart, 1993, pp. 236 ff

INDEX

ABOUT THE AUTHORS

Hans de Bruijn is professor of Organisation and Management at the Faculty of Technology, Policy and Management, Delft University of Technology.

Ernst ten Heuvelhof is professor of Policy Science at the Faculty of Technology, Policy and Management, Delft University of Technology and at the Department of Public Administration at Erasmus University Rotterdam

Roel in't Veld is dean of the Netherlands School of Public Administration, professor of Organisation Sciences at the University of Amsterdam, professor of Management of Public Organisations at the Utrecht School of Governance and professor of Hybrid Organisations at the Open University of the Netherlands.

Printed in the United Kingdom
by Lightning Source UK Ltd.
103028UKS00002B/57

9 781402 026805